Shoot the Moon

The Moon boasts an unexpected variety of landscapes, including volcanic features, sinuous valleys, and ghost craters readily visible from Earth. This practical guide explains and demonstrates how you can capture impressive images of our nearest neighbor in space using a variety of different techniques. As the first guide to be dedicated to modern lunar imaging, this volume offers an in-depth and illustrated approach to common optics, the essentials of digital images, imaging devices, and image-processing software. Even in light-polluted areas, the countless features and finest details of the Moon can be captured by following the instructions of this comprehensive and accessible guide. Covering equipment ranging from smartphones and DSLRs to specialist planetary cameras, whether you are a novice without a telescope, an amateur developing your skills in imaging or an experienced astrophotographer, you will benefit from the hints, insights, and expertise within.

Nicolas Dupont-Bloch is an amateur astronomer, based in Sautron, near Nantes, France. He has previously published two books (in French) and has translated a third into English. Many of his images have been showcased in *Astronomie Magazine* and chosen as the Lunar Picture of the Day (LPOD). This is his first English astrophotography guide.

Shoot the Moon
A Complete Guide to Lunar Imaging
Nicolas Dupont-Bloch

With image contributions from

Mikhail Abgarian (Belarus)
Mathias Barbarroux (France)
Gilles Boutin (Canada)
Dani Caxete (Spain)
Frédéric Géa (France)
Maxime Giraudet (France)
Yuri Goryachko (Belarus)
Etienne Martin (France)
Konstantin Morozov (Belarus)
Catherine Port (France)

All other photographs and drawings by the author
unless otherwise mentioned

Shoot the Moon

A Complete Guide to Lunar Imaging

Nicolas Dupont-Bloch

CAMBRIDGE
UNIVERSITY PRESS

University Printing House, Cambridge CB2 8BS, United Kingdom

One Liberty Plaza, 20th Floor, New York, NY 10006, USA

477 Williamstown Road, Port Melbourne, VIC 3207, Australia

314-321, 3rd Floor, Plot 3, Splendor Forum, Jasola District Centre, New Delhi - 110025, India

79 Anson Road, #06-04/06, Singapore 079906

Cambridge University Press is part of the University of Cambridge.

It furthers the University's mission by disseminating knowledge in the pursuit of education, learning and research at the highest international levels of excellence.

www.cambridge.org
Information on this title: www.cambridge.org/9781107548442

First published 2016

A catalogue record for this publication is available from the British Library

Library of Congress Cataloging in Publication data
Dupont-Bloch, Nicolas, 1963–
Shoot the moon : a complete guide to lunar imaging / Nicolas Dupont-Bloch.
Cambridge, United Kingdom : Cambridge University Press, 2016.
LCCN 2015046463 | ISBN 9781107548442 (paperback)
LCSH: Lunar photography. | Moon – Pictorial works.
LCC TR713 .D87 2016 | DDC 778.9/36991–dc23
LC record available at http://lccn.loc.gov/2015046463

ISBN 978-1-107-54844-2 Paperback

For Catherine, Léna, and Maud for their patience with regard to my astronomical obsessions

and

in remembrance of Pierre Bourge and Jean Lacroux for their talent and their independence.

Contents

Contents

Contents

Preface

Being bright, large, and easy to aim at, the Moon is probably the first celestial object we observe with our naked eyes, binoculars, or a telescope.[1] Usually, we move on to observing planets, stars, and then deep-sky objects. Most of us take a simple and occasional glance at lunar craters when we have no more objects to observe, no more astrophotograpy to do, after a nebula has set in the West and before a planet rises in the East. Furthermore, the Moon is very bright – a continuous spectrum, drowning the faint light of elusive galaxies. And, after having appreciated some dramatic craters and beautiful sharp mountain peaks, who could honestly be interested in these countless regular small craters, all these strange features whose names we cannot remember? At the same time, the Universe is a living, changing place: Jupiter shows ever-changing clouds, novae or comets can suddenly appear, and even numerous diffuse nebulae can be imaged from a light-polluted suburb with appropriate filters. That's why I didn't seriously look at the Moon for years – until I observed it through a small, professional telescope, during a cold night in town.

The image was incredibly sharp, with such a contrast that I suddenly understood that I could directly see the bare soil of another miniature "planet," with a variety of rocks, topographic features, grazing impacts, bright or dark areas showing the evolution and history of a whole world.[2] Of course, we all want to see smaller and smaller details on the Moon, such as craterlets, wrinkle ridges, domes, or channels, but large areas showing high contrast also provide major hints to understand the lunar soil: they show ancient lava fountains and blasts, the age of lunar "seas" (maria) and their mineralogical composition.

The Moon is actually an open book, telling the history of the whole Solar System to whoever wants to read it through a simple eyepiece. Geology, volcanism, fractures, impacts, giant cliffs, magnetic mysteries beneath strange surface features, overlapping or negative craters, curved mountains . . . Although there are no frontiers on the Moon, its differentiated landscapes clearly show that a history involving a huge variety of events has occurred, and the evidence remains there to be observed. Unfortunately, I was quite unable to understand this fascinating

---

[1] The Sun and solar eclipses are not mentioned in this book. Observing the Sun requires specially secured solar filters.

[2] Our Moon is one of the seven large moons in the Solar System. It is the only one, among these massive moons, which has an extremely sparse exosphere, very little water ice relative to the icy moons of Jupiter and Saturn, and hardly any volcanic activity. That is why the record of its geological history has remained unscathed, especially over the last three billion years.

story. With a moderate magnification of 166×, countless craters and other mineralogical or volcanic stories all appeared at once: this exceeded my capacity to carefully observe each detail.

However, imaging can record and emphasize these features, and this can far exceed the resolving power and contrast of a simple telescopic observation. That's why I undertook a photographic survey of the Moon, whereupon I realized that processing lunar images is a unique process, quite different from other types of imaging. I hope that the experience I have gained will help other amateur astronomers to successfully start and improve their lunar imaging under the best possible conditions, with a very wide range of equipment.

Our trek begins with a simple smartphone and a basic beginner's refractor.

1

Introducing lunar imaging

1.1 Seven ways to shoot the Moon

Amateur lunar photography can efficiently record vast areas or close-ups of the Moon with various instruments: basic and apochromatic refractors, different types of reflectors, video cameras, compact, bridge, and DSLR (digital single-lens reflex) cameras, camcorders, and webcams, as well as astronomical cameras, even smartphones. As the Moon is bright and not too difficult to aim at, it can be a perfect first step to beginning astrophotography. As an example, Figure 1.1 shows the author's very first lunar image, taken with a high-performance 200-mm (8-in) telescope, a good, classic film camera, and a fine photofilm.

After having gained some experience, it became obvious that any instrument and camera should be able to capture decent images of the Moon (Figure 1.2). The golden rule is to exploit the ways in which they perform best. All instruments have a more or less limited field of possibilities; some are excellent for specific subjects. The trick is to match the instrument with the right filter, the right camera, and the appropriate method to successfully shoot the Moon.

Figure 1.1 The author's very first lunar image, taken with a 200-mm (8-in) telescope and a film camera. It shows almost all possible flaws: bad framing, multiple reflections, overexposure, motion blurring, incorrect focusing . . .

Figure 1.2 Some years later, the author obtained decent close-ups of the Moon with some cheaper – although a little larger – equipment. The image is centered on 33° E, 68° S.

Experienced amateurs occasionally reach a 400-m resolution on the Moon, or even slightly better. No professional Earth-based telescope, with conventional film (or plates), was able to reach such a resolution. But modern telescopes fitted with digital cameras no longer image the Moon. This is because, for several decades now, professional planetology has relied mostly on interplanetary probes (along-side some of the greatest Earth-based telescopes with adaptive optics and orbital observatories studying today's targets such as Jupiter's moons). But amateur equipment and original techniques have evolved during the same period.

There are numerous ways to shoot the Moon. The observation site matters, and so does the aperture of the telescope, but *imagination* is the most important factor that gives rise to pleasing images. In this book, we will have in-depth, practical views of these techniques, how to fix some optical and mechanical flaws, and how to fit the optics to the camera (whatever kind it is, as summarized in Figures 1.3 and 1.4), and then we will learn how to process the images to unveil bright, tiny craterlets as well as charcoal-dark, ancient lava-fountain deposits. We shall admire the incredible variety of lunar landscapes, to learn about the history of the Solar System or, more simply, to scrutinize the closest extraterrestrial world.

We will see that a number of mechanical adapters can help to marry various optics, imaging devices, and mechanical parts.

- Smartphone (behind a small telescope, finderscope, spotting scope . . .).
- Camcorder (alone or behind a small telescope).
- Compact or hybrid still camera (alone or behind a telescope).

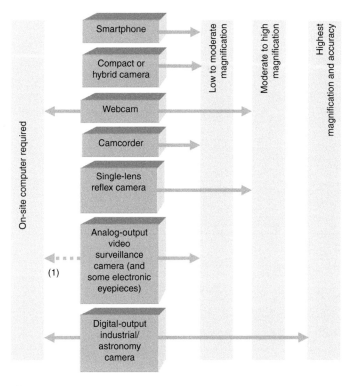

Figure 1.3 Common and specialized imaging devices and their possible scopes, whatever the optical equipment is. (1) Requires a video "grabber" (Section 2.13).

- Webcam (with a photolens or directly attached to a telescope with high magnification).
- Analog-output video camera with an analog-input monitor or a computer with a video grabber (directly attached to a telescope).
- Single-lens reflex camera (on its own or directly attached to a telescope).
- Digital-output industrial/astronomical camera (with a photolens or directly attached to a telescope).

This is now the second step of the debate. Which optics and combinations can be imagined, or are feasible, and when do they perform at their best? Any imaging device requires a lens. It can be included, removable or not. Many combinations are possible; e.g. a smartphone is self-sufficient but performs better with a telescope.

1.1.1 Beginners' equipment

We can start by holding a smartphone behind a small refractor (see Figures 1.5 and 1.6): the smartphone is simply put behind the eyepiece, instead of our eye. Star parties provide ideal opportunities to shoot the Moon for the first time.

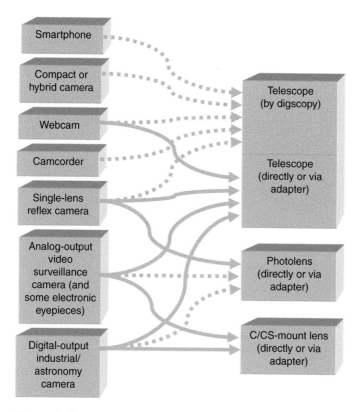

Figure 1.4 Combining imaging devices and optics. Dashed arrows indicate "plan B" or rare combinations.

A $30 webcam attached with the help of heavy-duty, opaque adhesive tape to a simple $90 refractor (Figure 1.7) shows surprising images of large craters, mountains, and maria. This is a reasonable expense to obtain our first lunar close-ups with a moderate magnification.

Many lunar subjects do not require a telescope: a lunar halo, a lunar eclipse, the basics of celestial mechanics illustrated by apparent movement variations of the Moon, a bright planet next to the Moon, the reflection of the Earth's sunlit side lighting up the dark side of the Moon, landscapes in moonlight, lunar phases . . . Most of these subjects are accessible to a common, $160 compact camera with a 60-mm zoom[1] or more (16× or 20× optical zoom). A more powerful bridge camera equipped with a 200-mm zoom (50×) provides wider possibilities in setting (especially manual focusing) and more detailed, smoother images. Lunar subjects need the use of a tripod, or you can place the camera on a low wall or at the edge of a

---

[1] We always mention the real optical characteristics, that is, the effective focal length of the photolens. Vendors often mention the equivalent 24 × 36 focal length, e.g. 250 mm in place of 60 mm. This is because sensors of compact cameras are smaller than those in DSLRs, leading to this equivalence.

Figure 1.6 The image is processed to gently enhance contrast and sharpness. The Moon is upside-down as viewed from the northern hemisphere through numerous astronomical instruments. This was the only rescued image out of twelve blurry, fuzzy, cropped, and overexposed still images.

Figure 1.5 Shooting with a smartphone held by hand behind a small refractor equipped with an 8-mm eyepiece. The mount was undriven (like a simple, sturdy photo tripod). Image by Catherine Port.

table, orientating the camera by placing it on a piece of folded fabric. The installation has to be reasonably firmly secured by keeping the strap around one's wrist; heavy-duty adhesive tape is also useful in many cases.

1.1.2 Intermediate equipment

This means a "standard," portable telescope: a 90–200-mm (3.5–8-in)-diameter reflector (an example is shown in Figures 1.8 and 1.9), an 80–110-mm (3–4-in) ED/apochromatic refractor, or a 100–150-mm (4–6-in) achromatic refractor. Such equipment performs well for earthshines, conjunctions, and lunar eclipses with a rather large field of view and a moderate to high magnification. An experienced astrophotographer may have to spend $550 to $2500 for the whole setup, not

Figure 1.7 The housing of the webcam is partially removed to unscrew its lens (the plastic housing is fragile and opening it voids the guarantee). Then the webcam, with its bare sensor, is attached to a small refractor with no eyepiece with the help of heavy-duty, opaque adhesive tape. The refractor is placed on a stable video tripod, allowing you to acquire decent still images, although the tripod cannot compensate for the diurnal motion.

including the computer. The ease of use and the total weight and bulk matter as much as the price, and the best instrument is the one we can use with pleasure.

Shooting lunar close-ups requires the following:

- a debut "planetary" camera ($70–110) or a better-performing, monochrome planetary camera (starting from $230);
- or a DSLR ($400, not including the lens);
- or a webcam (for cheapskate astrophotographers, $25);
- a laptop or desktop able to perform mid- to high-speed image capture ($450), even if it doesn't do so well at heavy image processing, especially stacking (the latter may be batch-executed, while we are sleeping);
- a fine, 127–200-mm reflecting telescope (starting from $425);
- and its motorized mount (starting from $470) with its tripod, to cancel the apparent motion of the Moon and to facilitate the orientation of the telescope because the camera has a very narrow field of view.

The mount is sold separately or with the optical tube. A German-type equatorial mount is better suited to astrophotography for all subjects, and it can be useful even without a telescope, for instance to shoot lunar eclipses with a camera and a telelens.

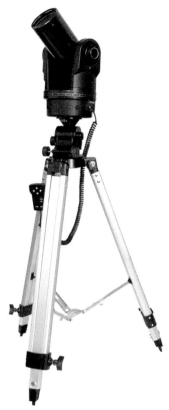

Figure 1.8 A widely used, intermediate-level telescope. With a compact design, a light weight, a motorized mount, and stable, good optics, it delivers very sharp images given the aperture.

Figure 1.9 This image is from a processed movie taken with a planetary camera directly inserted into the eyepiece holder of the compact, 90-mm motorized telescope shown in Figure 1.8. Given the long focal length of this kind of reflector, known as a Maksutov–Cassegrain reflector, along with the tiny photosites of the camera, the direct adaptation offers appreciable magnification and pretty good sharpness. Despite its undeniable optical quality, the magnification of the telescope is limited by its diameter (41° E, 30° S).

Bigger telescopes perform better, but they are more expensive, cumbersome, and heavy. In addition to this, they are more sensitive to turbulence and wind; their mirrors need to be controlled often (taking some tens of seconds) and possibly aligned (taking a few minutes, with experience). They have typically 180–250-mm apertures, and they weigh 30–50 kg (they can be broken down into components weighing 2–15 kg).

A telescope is not mandatory; numerous subjects need the following:

- a DSLR with a good 300-mm photolens ($1700) or a good compact or hybrid camera ($300+);
- a sturdy photo/video tripod ($220);
- or, for taking images or movies of lunar eclipses, a mount with one motorized axis and a counterweight, for loads of up to 2 kg ($300) or up to 8 kg ($550 with no GOTO[2]).

1.1.3 Advanced equipment

An experienced astrophotographer may purchase more expensive equipment:

- a high-end planetary camera (typically $450–2800);
- a 15–30 kg, 280–350-mm (11–14-in) catadioptric OTA (optical tube assembly) ($2500–8000),
- or a 250–400-mm (10–16-in) Newtonian telescope ($800–6000)
- and its motorized, sturdy mount ($1800–4700 for lunar imaging[3]),
- or, alternatively, a 50–80-kg, 400–600-mm (16–24-in) collapsible and motorized Dobsonian telescope ($3800–10 000);
- various accessories (filters, correctors . . . $350–1700) with an optional car battery;
- a reasonably good laptop or desktop (about $800) for high-speed image acquisition and processing;
- and it helps to own a backyard in a dry, high-altitude desert with a permanently steady atmosphere.

This is not entirely true. Experience shows that affordable 250–300-mm (10–12-in) telescopes (Figure 1.10) offer a good mean diameter to achieve high-resolution imaging. Although they have a limited resolution, they are less sensitive to wind and turbulence than are telescopes with larger diameters: some tens of nights of good atmospheric conditions per year can be exploited. Telescopes of diameter 350 mm and larger deliver their best lunar images about five nights per year from common locations (in the suburbs, in the countryside, or at the seaside).

This may seem surprising, but high-end, high-cost, apochromatic refractors are not often mentioned in this book. This is because, despite their image quality indeed being magnificent, the average diameter seldom exceeds 120 mm (180-mm and even bigger apochromatic lenses exist . . . at an astronomical cost), resulting in a relatively moderate resolving power; so they cannot compete with low-cost, large-aperture telescopes when the atmosphere is calm. The author exploited a 150-mm

---

[2] A computerized mount, able to automatically aim at ("go to") a celestial object after an initialization step. A motorized mount does not require a GOTO facility to track the Moon. This functionality may be disengaged for the sake of simplicity.

[3] Planetary and lunar imaging do not require a very stable mount for long exposures, nor do they require precise sidereal tracking, autoguiding, or GOTO.

Figure 1.10 A 200–300-mm (8–12-in) Newtonian telescope or a 250–280-mm (10–11-in) Schmidt–Cassegrain telescope resolves 600-m details. This is a portable but cumbersome and heavy instrument, typically weighing about 30–50 kg (66–110 lb).

(6-in), F/D = 8, triplet apochromatic refractor, but a low-cost (fourteen times cheaper), conveniently prepared, 250-mm (10-in) Newtonian telescope proved to be more adapted to lunar imaging when the atmosphere is steady.

Copernicus through a one-meter, DIY Dobsonian

Frédéric Géa (France) imaged the Copernicus crater (Figure 1.11) with the unusual Dobsonian he had built (Figure 1.12):

"For a long time, I thought that the Moon existed only to prevent me from deep-sky observing. In March 2013, with the hereafter motorized 1000-mm (40″) Dobsonian and a DMK31 camera with a near-infrared, 742-nm filter and a 3× Barlow lens, despite a demanding focusing because of the F/D = 3 prime focus, the image of Copernicus seemed steady and I captured some movies. On the next day, the friend I asked to process the stacked images (I was a beginner in image processing) told me, with a strange voice, that they had a *certain potential.*"

Figure 1.11 "Observing the Moon with a one-meter telescope is fascinating, especially with a binoviewer, albeit dazzling. When the atmosphere is calm, the show is splendid, with a feeling of relief. Depending on the turbulence, tiny craterlets appear or vanish. The extraordinary subtlety of shades of this powdery soil, like a gray talc, is surprising, and sometimes I'm almost expecting to discover the tracks of imaginary creatures." (Frédéric Géa).

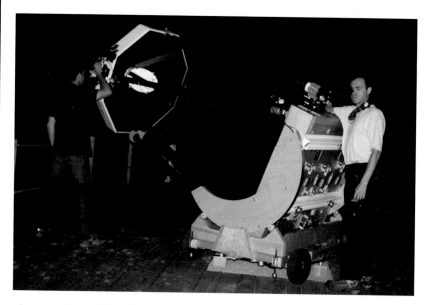

Figure 1.12 Frédéric (on the right).

Frédéric's blog: www.stellarzac.com/eng/.

1.2 The Moon's changing appearance

1.2.1 Monthly cycles and proper motion

Being a natural satellite, the Moon orbits its primary, the Earth. One revolution takes 27.3 days; this is the time interval for the Moon to pass again in front of the same star as viewed from the Earth. This is the sidereal period, because it refers to a distant star. During the sidereal period, the angle at which we see the lit part of the Moon changes; this is why we can see phases on the Moon, like on Venus, and even on Mars and Jupiter. Of course, while the Moon orbits the Earth, it follows the Earth traveling around the Sun. When we look carefully at lunar phases, this leads us to define another important duration of 29.5 days (e.g. from a full Moon to the next full Moon). This duration is the synodic period: the time interval for the Sun, the Earth, and the Moon to be aligned again. The synodic period determines the visibility of different lunar features, the hours at which it rises and sets, and the duration of its visibility – it also depends on the season and the location of the observing site. For instance, a few days after the new Moon at the latitude of 45° N on August 29, 2014, the thin crescent is only 14° high (the size of our hand, thumb extended, with the arm outstretched) at sunset, then the Moon sets ninety minutes later. It is not recommended to try to image the Moon in daylight because this strongly decreases contrast – not to mention the danger of solar light in the vicinity.[4] The solar light can definitively damage the camera, the secondary mirror of a reflecting telescope, or a catadioptric photolens, the cross-hair in the finder, and ... our eyes. On the contrary, we can take advantage of the fact that the full Moon is visible all night long, more or less. A lunar imaging session can be casual and simple, but advanced imaging requires a bit of preparation with the help of ephemerides in astronomy magazines or specialized software and freeware (Appendix 2).

Another consequence of the lunar orbit is the slight, but not negligible, difference of 0.5 arcsecond (1/7200 of a degree) per second between the diurnal motion and the apparent motion of the Moon. This difference, the proper motion of the Moon, has to be compensated for when we need long exposures, for instance at high magnification with a moderately sensitive sensor, or during the darkest step of a total, lunar eclipse. This is why numerous motorized mounts offer a particular "lunar rate," "lunar tracking," or "lunar speed" to accurately follow the Moon. Nonetheless, standard sidereal tracking is sufficient for lunar imaging, possibly with slight, manual corrections with the help of the hand controller.

---

[4] A red or near-infrared filter strongly dims the daylight, at the price of an extended exposure duration. This demands a low-noise camera, a motorized mount, and the taking of extreme precautions because of the close proximity of the Sun. The Sun has to be permanently masked out by a wall or a secured shield.

1.2.2 Apparent diameter variation

The Moon has an elliptical orbit around the Earth. With a high eccentricity of 0.055, which is greater than that of all other major moons such as Ganymede and Titan, its distance from us varies from 356 300 km to 406 700 km (center to center). As a consequence, the lunar diameter shows some variation: approximately 7% (and a variation of 30% in brightness). When the full Moon is at its very closest point, the situation is popularly known as a "supermoon" – the scientific expressions are perigee-syzygy and perigee Moon. When we image the Moon at its closest approach, we have the benefit of a larger angular size, or, in other words, free additional magnification. This actual variation is not to be confused with the apparent (optically illusory) change of its diameter when the Moon is low above the horizon. The most interesting supermoons occur at full Moon, about 4–6 times per year (half of the supermoons are new Moons). The next most spectacular ones will happen on the following dates:

November 14, 2016
January 2, 2018
January 21, 2023
November 25, 2034
January 13, 2036

Less spectacular supermoons occur more frequently. Fred Espenak maintains a list of supermoons at http://astropixels.com/ephemeris/moon/fullperigee2001. html.

Most years encompass three or four supermoons, even if they do not reach the maximal size of those on the dates mentioned above.

1.2.3 Visibility of lunar features with respect to phase

Lunar contrast, the difference between dark and bright areas, is strong but evolving because of the lunar day; it also decreases at high magnifications.[5] Topographic features are revealed by the shadows cast, that is, close to the terminator where the light is incident at a low angle. Contrast is poor far from the terminator and on approaching the full Moon: since the illumination is more or less vertical, the brightness is strong but there are no shadows cast (see Table 1.1). On the other hand, as we will see later, differences in reflectivity due to the composition and age of the soil are emphasized. The solar wind, micrometeorites, and mass wasting act on the reflectivity with time: the soil becomes darker, the

---

[5] Apochromatic refractors maintain an excellent contrast even at high magnification because of their optical quality and their light transmission ratio of 98%–99%, to be compared with 84% for most reflectors (about 92% for each mirror, with the exception of some Ritchey-Chrétien devices, whose mirrors have a reflectivity of 99% due to dielectric coating), in addition to the difference of precision in optics grinding.

Table 1.1 The illumination angle, which is directly related to the phase, determines the visibility of different lunar features. A high illumination angle (full Moon) unveils differences in soil composition and age. A low-angle illumination casts elongated shadows and reveals shallow and low-altitude reliefs.

	Crescent	Quarter	Gibbous	Full
Contrast of maria, pyroclastics, ray systems	Poor	Good	Very good	Excellent
Visibility of relief at terminator	Excellent	Excellent	Excellent	Nil

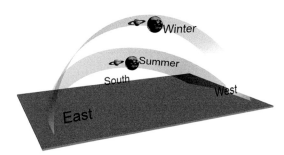

Figure 1.13 The apparent height of the Moon (and main planets) varies with the season. This drawing shows the nocturnal, seasonal variation viewed from the northern hemisphere; seasons are inverted for the southern hemisphere (or the reverse).

surface becomes more rugged, and pebbles fall down along slopes. After billions of years (namely the last 3.5 billion years), the lunar soil tends to be less reflective. Today, the reflectivity is no more than 5%–15%: indeed, the Moon is a mediocre mirror. The global albedo – the ratio between the reflected solar light and the incoming solar light – is 12%, compared with the mean terrestrial albedo of 39%, thanks to oceans, polar caps, and clouds.

1.2.4 Seasonal cycle and altitude variation

Because of the inclination of the rotation axis of the Earth (23.5°) and the inclination of the orbital plane of the Moon (5.1°), the height of the Moon varies by 36–56° at midnight.[6] For instance, as illustrated in Figure 1.13, when the Moon is full during winter in the northern hemisphere at the latitude of 48° N, it rises from the North-east, reaches its maximum altitude when it passes South at midnight (Universal Time), and then sets to the South-west. In December, the full Moon is at 57° above the horizon (almost three times the width of our hand at right angles to the horizon, with arm outstretched and fingers spread wide). Conditions are perfect to image the full Moon. But in June, viewed from the same location, the Moon barely reaches 21°. It seems yellowish, and the turbulence is

---

[6] See www.nfo.edu.moonview.htm and "lunar standstill" at wikipedia.

substantial because the image of the Moon passes through more air. The scene may be pleasant, but the situation is not favorable for viewing at high magnifications.

The best thing to do is to image the Moon when it is high, that is,

- for the northern hemisphere, during the last quarter of Fall and the first quarter of Winter;
- for the southern hemisphere, during the last quarter of Spring and the first quarter of Fall.

Equatorial locations are favored considerably because the Moon rises very high all year long, crossing less air, and hence imaging it is less prone to turbulence and atmospheric refraction.

Useful tools to plan the imaging sessions are the ephemerides in astronomy magazines and planetarium software (Appendix 2).

1.2.5 Near-side visibility and librations

The Moon orbits the Earth according to the same laws as a planet around its star. Like many moons in the Solar System, it has an unevenly distributed mass, so it is more attracted by terrestrial gravity where it is locally more massive: beneath the maria.[7] They are numerous on the "heaviest" side of the Moon, which is always facing us: the near side.

Because of the pronounced orbital ellipticity, both distance and orbital speed vary. At apogee (the greatest distance from the Earth) the Moon travels along its orbit at a speed of 0.96 km per second; at perigee (when it is at its closest to the Earth) it speeds up to 1.07 km per second. Hence, there is a slight horizontal angular variation between the Earth and the Moon, and we can see a little more than half the Moon in width: this is the main variation of its visible side, or main libration. As the Earth rotates (diurnal motion), another subtle variation of the angle brings an additional libration.

Another non-negligible libration is due to the 5.15° inclination of the lunar orbital plane to the ecliptic (the orbital plane of the Earth around the Sun) and to the 1.5° inclination of the lunar polar axis with respect to its orbit around the Earth. This 1.5° polar inclination is added to or subtracted from the 5.15° orbital-plane inclination. Thus, the lunar poles seem to swing a little during the lunar revolution.

The combined effects of librations, with a total amount of about 7° in both vertical and horizontal axes,[8] allow us to see 59% of the Moon's surface. Marginal

---

[7] These subtle local gravity anomalies are strong enough to lower the orbit of lunar probes. This even offers an efficient way to remotely study the subsurface of planets and other moons. See a lunar gravity map at https://solarsystem.nasa.gov/multimedia/display.cfm?IM_ID=8024.

[8] 7° 54 arcminutes in longitude, 6° 50 arcminutes in latitude.

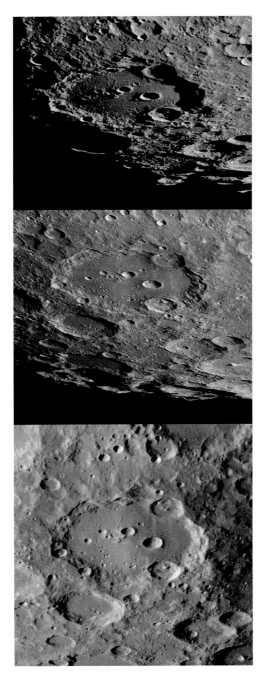

Figure 1.14 Effect of polar libration (14° W, 58° S). The equatorial libration is illustrated in Figure 11.29.

features can appear larger on some dates, or may be horizontally flattened due to perspective. Horizontal libration is the most conspicuous effect, but polar libration can help too, as illustrated in Figure 1.14.

15

1.2.6 Colongitude

The colongitude is the longitude of the terminator, that is the location of the separation between the lunar day and night. As the Moon rotates around the Earth with respect to the Sun, the terminator slides westward by 0.5° per hour ("westward" stands here for selenographic coordinates, lunar West is on our left hand if we observe the Moon from the northern hemisphere). The colongitude is 0° when we see a quarter Moon, 90° when we have a full Moon, and so on. The difference between the longitude of a relief and the colongitude is interesting because it is directly related to the elongation of the cast shadows. It can be calculated precisely by software or it can be estimated by subtracting the longitude (fixed) of a crater from the approximate colongitude (variable, see the ephemerides in magazines or lunar software in Appendix 2). Reliefs are best visible when the difference is small, or, in other words, when they are close to the terminator, with less than a 15° difference in colongitude. Many astrophotographers are attracted by shooting the craters only when they are close to the terminator, because the relief is conspicuous, but many other features with no relief (e.g. maria, ray systems, pyroclastics) are visible when the difference in longitude is more than 30°.

2

Choosing your imaging equipment

2.1 Shared, fundamental characteristics

Photolenses, refractors, and reflectors share two basic characteristics and a derived one.

- The focal length, or optical distance between the objective or the main mirror and the image sensor; this determines the size of the image. This is not the most important factor because this magnification can easily be increased or decreased as long as it is not excessive with respect to the diameter and the optical quality.
- The diameter of the objective or of the main mirror; this determines the resolving power (accuracy), the contrast, and the sensitivity to turbulence.
- The focal-length/diameter ratio, or F/D ratio, an important parameter for photography; it determines the exposure duration, the ease of focusing, and some geometrical characteristics which affect the image field.

A golden rule is that, when the focal length is large with respect to the diameter, optical flaws are minimized. As the diameter increases, the F/D ratio of an achromatic refractor has to increase if we are to maintain a comparable quality (e.g. from F/D = 8 for an 80-mm (3-in) achromatic refractor up to F/D = 15 for a 150-mm (6-in) achromatic refractor). Reflectors and apochromatic refractors are less affected. Furthermore, devices with "slow" optics (when the F/D ratio is more than 8 or so) are cheaper because they are easier to manufacture while maintaining the same – or better – optical quality, easier optical adjustment, and possibly a better ease of use even with a basic focuser. Lunar imaging, unlike most astrophotographical subjects, does not demand a "fast" (<6) F/D ratio.

2.2 Optical flaws

Chromatism and astigmatism affect the entire image. Other optical deformations affect mainly the corners of the image.

- Coma transforms points at the corners into little comet-like spots.
- Distortion either stretches the corners or inflates the center.

- Astigmatism transforms points into crosses when the image is focused, or vertical or horizontal lines when it is out of focus.
- Chromatism is the impossibility for the optics to make all colors converge at the same place. While the image is sharp in red, it is somewhat unfocused in blue. In addition, lenses suffering from chromatism also focus the blue poorly and waste light.
- Vignetting dims the corners of the image.
- The optics' grinding and polishing quality affect sharpness and contrast.
- Misalignment of the optics affects sharpness.
- In some cases (especially digiscopy), the field is curved, so that focusing is possible only at the center or at the corners.

Some types of flaws are mainly visible at the corners of the image. That is why a "full-frame" DSLR (24 × 36-mm sensor) requires high-end optics and a field corrector to get a perfect image over the entire field, while the small sensor of a "planetary" camera acquires only the very center of the image field, where most optical flaws are diminished or negligible. Figure 2.1 shows a defocused, single star centered in the field.

These flaws can be assessed by taking an image of a field of bright, bluish stars, for example the Pleiades. Some can be directly observed in an eyepiece, but they result from the combination of the main optics and the eyepiece optics, so such a test could not be representative of the real imaging field. The most visible flaws are astigmatism, effects of misaligned optics, and chromatism. We will see in Chapter 4 how to fix some flaws.

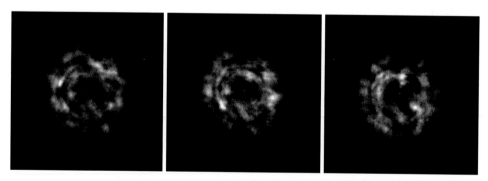

Figure 2.1 A strongly magnified star, deliberately defocused, in a Newtonian telescope. The differences in the three frames are due to the turbulence. Some shadows stand out in the diffraction pattern, due to various factors: clipping of the primary mirror, spider vanes, the drawtube being in the light path, and roughness of the optics. If the pattern is elongated when the star is in focus and elongated at right angles when it is out of focus, there is astigmatism. If the diffraction rings which are respectively in and out of focus show different distributions of light, the optics will suffer from spherical aberration (a very common flaw), causing imperfect focusing.

2.3 Resolving power

One of the most important characteristics for a good lunar telescope is the resolving power. Since lunar craters are more contrasted than jovian clouds, we do not need cutting-edge optics to start lunar imaging. We need merely diffraction-limited optics, that is, a telescope able to actually distinguish two close points (see Figure 2.2) with respect to its theoretical possibilities. The theoretical resolving power is calculated with the help of the Rayleigh criterion:

$$\text{Resolving power} = 1.22 \times \text{Wavelength/Diameter}.$$

The wavelength is the physical characteristic of the color of the light.[1] With a 203-mm (0.203-m) telescope in green light (of wavelength 0.000 000 56 m), the result is

$$
\begin{aligned}
\text{Resolving power} &= 1.22 \times 0.000\ 000\ 56/0.203 \\
&= 0.000\ 003\ 365 \text{ radian} \\
&\quad (\text{conversion from radians to degrees is radians} \times 180/3.14) \\
&= 0.000\ 192 \text{ degree} \\
&\quad (\text{conversion from degrees to arcseconds is degrees} \times 3\ 600) \\
&= 0.69 \text{ arcsecond, about } 1.2 \text{ kilometer on the Moon for observation.}
\end{aligned}
$$

This is the most commonly used criterion to rate the resolving power for observing (other variants exist, such as Dawes' criteria and Sparrow's criteria, for different colors).

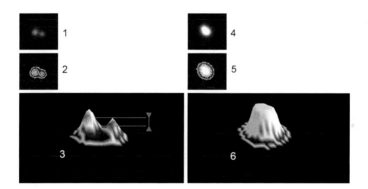

Figure 2.2 Top left: a real image of two close stars. Owing to various factors,[2] stars are not pinpoints in a telescope. Middle left: isophotes show that, despite the stars being resolved (separated), they also share a common surface of low brightness levels. Bottom left: a 3D representation of the levels. The difference (arrow) is the contrast (actually the measurements are performed at half the height). Right: an unresolved pair of stars. In addition, overexposure clips the levels. In lunar imaging, resolved but clipped areas like bright crater rims merge into an undifferentiated spot. Thus, lunar imaging requires a good angular separation, good contrast, and correct exposures.

[1] It is measured in nanometers (1 nm = 0.000 000 001 m) or in ångström units (1 Å = 10 nm).
[2] Diffraction forms Airy disks, that is, concentric rings with a bright central disk, while turbulence spreads light, especially during long exposures, along with optical surface roughness and other optical flaws.

19

In practice, we have the following criteria.

- The **resolving power** is the tiniest angle the optics can distinguish. It is calculated in radians, a necessary though non-intuitive unit, but values in radians can easily be converted into degrees or arcseconds, which are more useful units for lunar and planetary imaging. A whole circle – a panoramic view when we complete one turn around our vertical axis – is 360 degrees. Each degree comprises 60 arcminutes, each of 60 arcseconds. The full Moon is roughly 30 arcminutes wide, tens of times larger than planets. The actual resolving power of a good 200-mm (8-in) telescope is about 1 arcsecond or a little less, theoretically 0.69 arcsecond according to the Rayleigh criterion. Nonetheless, good and conveniently processed images obtained with such a telescope may show a resulting resolution of 0.35 arcsecond. The Hubble Space Telescope can reach a 0.1–0.05-arcsecond resolution with a 2.4-m mirror and, obviously, no turbulence.

- The **wavelength**, often given the symbol λ (lambda), is the color of incoming light. Since the Moon is roughly gray, we could equally favor any precise wavelength, for instance a yellowish green at 560 nm, which is the center of the best sensitivity of the human eye. Or we could choose the best-focused color of the instrument, or the color for which turbulence is less destructive. For example, a near-infrared image can be taken around 850 nm to diminish the effect of turbulence, but our 203-mm telescope will suffer from this larger wavelength with a lowered resolution of 1.07 arcsecond, that is a 33% accuracy loss with respect to visible light.

- The **diameter** is the aperture of the telescope, in millimeters (mm) or inches (in). Its influence is shown in Figure 2.3.

Table 2.1 shows the limits to visual resolution with a steady atmosphere. The resolution is often worse in windy or unstable temperature conditions. But, when the atmosphere is calm, imaging techniques can exceed the limit by a factor of 2.5. This is equivalent to deep-sky photography unveiling countless stars in minutes while they remain invisible to the unaided eye.

2.4 Photolenses and zooms

A photolens with a short focal length (9–135 mm) is perfect to image a landscape with the Moon, lunar eclipses, earthshines, conjunctions, and wide-field subjects. A photolens with a long focal length (300 mm) can show the entire Moon with numerous craters and mountains. Unlike for deep-sky imaging, a "quick" and expensive photolens is not necessary ("quick" refers to a low F/D ratio, e.g. 2.8). The major optical flaws of an ordinary telelens can be ameliorated by closing the iris by two notches or much more. We will often use zooms at their maximal focal length, where they are less prone to optical flaws. One can buy used fixed-focal-length photolenses from classic film cameras at low prices, especially those fitted

Table 2.1 Resolving power for observation or basic imaging. The limits have been superseded by modern amateur imaging.

Diameter of optics		Resolving power (arcsecond) relative to wavelength			Finest detail on Moon (observation)	
		Blue,	Green, 560 nm	Red,	Near infrared.	
mm	in	430 nm	(human vision)	650 nm	740 nm	(km)
508	20.0	0.21	0.28	0.32	0.37	0.5
457	18.0	0.24	0.31	0.36	0.41	0.5
356	14.0	0.30	0.40	0.46	0.52	0.7
306	12.0	0.35	0.46	0.53	0.61	0.8
254	10.0	0.43	0.55	0.64	0.73	1.0
203	8.0	0.53	0.69	0.81	0.92	1.2
178	7.0	0.61	0.79	0.92	1.05	1.4
152	6.0	0.71	0.93	1.08	1.23	1.6
127	5.0	0.85	1.11	1.29	1.47	1.9
102	4.0	1.06	1.38	1.60	1.83	2.4
90	3.5	1.20	1.57	1.82	2.07	2.7
80	3.1	1.35	1.76	2.04	2.33	3.1
63	2.5	1.72	2.24	2.60	2.96	3.9

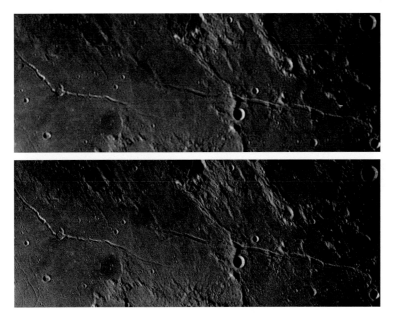

Figure 2.3 The aperture of a telescope conditions the contrast and the resolution. These two cropped images were taken on the same night some kilometers apart. Top: 125-mm (5-in) reflecting telescope. Bottom: 254-mm (10-in) telescope. The small telescope is limited by its diameter but it acquires constant-quality images all year long, whereas the larger one is prone to turbulence blurring (11° E, 7° N). Top image by Etienne Martin.

to Nikon, Canon, and Pentax cameras, because they have readily available mechanical adapters to connect them to an astronomical camera.

A catadioptric telelens comprises both mirrors and lenses; it often has a fixed focal length of 500–1200 mm (Mangin and Maksutov types). The image quality can be very good. The secondary mirror (the opaque disk at the center of the front meniscus) has to be as small as possible, indicating that the optics are not designed to be as quick as possible (F/D = 5–6), but to be as fine as possible (F/D = 8–11).

A modern photolens can be said to have been especially designed for digital cameras. There are two real differences from an old photolens.

(1) The presence of an internal, anti-reflective coating to avoid reflection of light between the shiny, glass protection window above the sensor and the last internal lens of the photolens. This can be a problem because the Moon is very bright, and the protection window of a sensor is much more reflective than a classic film support like ESTAR. A bright image reflects to the internal lens, and then reaches back to the sensor, possibly causing ghost images.
(2) A shorter focal length for small-sensor cameras, especially in compact still cameras. That is why the focal length may be indicated in two ways:
   - the actual focal length (in millimeters);
   - the equivalent focal length (in millimeters), that is, the focal length the lens should have in order to provide the same magnification if it were mounted on a full-frame (24 × 36-mm-sensor) DSLR.

The actual focal length is always indicated in front of the lens, as illustrated in Figure 2.4. A good thing is that an old photolens – designed for classic 24 × 36-mm format – exploited with a small sensor casts only the center of the image, which always shows attenuated optical flaws, while offering some free extra

Figure 2.4 Left: a vintage, fixed-focal, 135-mm, F/D = 2.8 telelens. Right: a modern, variable-focal-length, 18–70-mm, F/D = 3.5–4.5 zoom. Both lenses display their F/D ratio and actual focal length, engraved at the front or rear of the body. The F/D ratio of a zoom varies while the focal length is adjusted.[3] Both can equip an astronomical camera or a modern DSLR, at the price of manual focusing, use of a manual iris, and there being no help at all for automatic exposure.

[3] Some high-end zooms maintain a constant F/D ratio.

magnification. With a 15×22-mm APS-C-format camera, the magnification is about 1.6×. This changes when the lens is fitted to an astronomical camera because sensors exist in a wide variety of formats.

2.5 Refractors for beginners

A very affordable, debut 60–70-mm achromatic refractor (Figure 2.5) on a simple azimuthal mount allows imaging with a smartphone or a compact still camera with digiscopy or, for better results, with an astronomical camera, a webcam, or a DSLR at prime focus. The optical quality and the generally unstable, unmotorized mount limit use to short exposures and low magnifications, but one can shoot craters. No image processing could noticeably improve the images, with the exception of a moderate contrast enhancement. No close-up at high magnification is possible.

The main drawbacks are chromatism (actually spherochromatism), coma, and field curvature, especially if the "achromatic" refractor has a low F/D ratio (5 or 6).[4] Chromatism tints shadows in blue–purple and the lunar surface is greenish or yellowish, but this can be fixed. The flaws are attenuated by using a small sensor and, if the refractor has an F/D ratio of 8 or more, the resolution appears rather good, albeit with poor contrast.

Figure 2.5 A debut refractor comprises an objective made of two lenses, baffling to reduce internal reflections, and a focuser. The imager can be placed in the drawtube with various assemblies (Chapter 3) to adapt the magnification.

---

[4] A "rich-field," F/D = 5 achromatic refractor is designed to allow us to admire stellar fields and deep-sky objects at low magnification.

2.6 Advanced refractors

Chromatism is the main flaw of refractors for lunar imaging. The only way to almost completely, or totally, suppress chromatism is to use "apochromatic" refractors. Their objective comprises two or three lenses, one of which is made of a special type of glass containing rare-earth elements to offer very low differential dispersion of the colors: an ED ("extra-low dispersion") glass or, more rarely, FLUO glass (containing fluorite). Such lenses are "semi-apochromatic" (generally two lenses) or "apochromatic" (generally three lenses), and colors converge to the same point, resulting in a contrasted, bright image, with natural colors and crisp details, while achromatic refractors provide sharp images in one color only.

One or two lenses are made of classic "crown" glass (e.g. Schott BK7), and the apochromatic or ED element is made of special glass, characterized by an important parameter called the Abbe number, which specifies the variation of light refraction with respect to the wavelength. The higher it is, the better the special glass. Here are some glasses and their Abbe numbers:

- N-BK7 (crown):   64.20
- FPL-51 (ED):   81.54
- H-FK61 (ED):   81.61
- FCD1 (ED):   81.61
- OK-4: (ED):   92.1
- FPL-53 (ED):   94.93
- CaF_2 (fluorite):   94.99

At its highest practical magnification (180× with a Nagler eyepiece adapted to a "quick" F/D ratio), an affordable and popular 80-mm refractor with an FPL-53 ED doublet (two lenses) and a classic F/D = 7 ratio shows very little chromatism, though images of the brightest stars show a tiny, bluish halo. A 152-mm (6-in), OK-4 triplet (three lenses) with a reasonable F/D = 8 ratio shows neither chromatism nor halo, and magnifications of 300× are quite common. The best refractors use triplets with one OK-4 or CaF_2 lens. Some FPL-53 doublet refractors with a reasonable F/D = 8 ratio are very similar to CaF_2/OK4 refractors. Even an FPL-51 doublet refractor is far better than any achromatic refractor: despite a residual, blue halo around bright stars and a small amount of greenish/yellowish tint, the light is correctly concentrated, with excellent contrast.

An objective is not made solely of glass. Frequently, spaces between elements are filled with air, or sometimes oil, to minimize losses due to air/glass interfaces and reflections. Lenses also require precise grinding and polishing, with addition of anti-reflective coating, and must be conveniently placed in an adjustable cell. The only way to determine precisely the quality of an objective is to perform optical tests. The results are difficult to interpret for those of us who are not opticians. Many optical tests on commercial products have been conducted by Wolfgang Rohr:

http://r2.astro-foren.com/index.php/de/berichte/01-aeltere-berichte-auf-rohr-aiax-de-alles-ueber-apos

These web pages are in German, but they are extremely interesting and include figures and screen captures in English.

Three-lens objectives are often best, but they have three drawbacks:

- aligning three lenses is an impossible task for an amateur;
- loosening the cell to align oil-spaced optics is not a good idea at all;
- a three-lens objective is really heavy and the instrument needs a sturdy mount, with a high tripod because the main mass is in front of the optical tube, thus the tube must be placed very far back to achieve a proper balance (except for refractors equipped with a heavy focuser).

Apochromatic refractors can have more than three lenses: four- and five-lens formulas have a two- or three-lens objective plus an internally placed field corrector (which is often simultaneously a focal-length reducer). Such refractors are superbly designed for deep-sky imaging. The photographic field of view is planar and flawless, more or less, for the large sensors of DSLRs and deep-sky cameras. But we have to consider the following points.

(1) Good sensors for lunar imaging are often of size 1/2 in or smaller,[5] so they do not need a large, corrected photographic field.
(2) The corrector often enlarges the smallest points the objective alone can produce – but the photographic field has a homogeneous quality.
(3) A number of great refractors can be purchased with or without a corrector/focal-length reducer. Without a corrector, the image is less corrected in the corners, but it may be more accurate in the center. In addition, the refractor keeps its genuine focal length: this means less further magnification is needed for imaging, and this preserves a good contrast.

Finally, an old rule remains unchanged: given an aperture, the longer the focal length, the more flawless the image field. This means that a "slow" F/D ratio, let's say 8, guarantees an excellent refractor for planets, and it may be exploited later with an optional corrector/reducer for deep-sky imaging. And, being refractors, ED/apochromatic refractors have no spider, no secondary mirror ... The images are undoubtly purer than those obtained in reflectors.

So, why are apochromatic refractors not the "standard" means to image the Moon? As mentioned previously, the resolving power depends primarily on the aperture. An excellent 150-mm apochromatic refractor costs $14 000 just for the optical-tube assembly (OTA). This is the price for twelve well-performing 200-mm F/D = 6 Newtonian OTAs optimized for planets and the Moon. And they have

---

[5] Some industrial sensors are quite large, e.g. CMOSIS CMV2000 (2/3 in) and CMV4000 (1 in).

roughly the same resolving power, even if apochromatic refractors always beat Newtonians in terms of overall image quality, contrast, stability, and repeatability.

2.7   Reflectors and catadioptric telescopes

Pure reflectors comprise mirrors only (Table 2.2). The most affordable, though still well-performing, is the Newtonian telescope (as in Figure 2.6); unfortunately, it has numerous drawbacks such as its weight, bulk, sensitivity to wind and internal turbulence, and the necessity to periodically align the two mirrors. Other reflectors may be folded (see Table 2.3) to limit their bulk. Unlike in Newtonian reflectors, the secondary mirror is not planar: it participates in the field correction and the magnification.

It is wise to avoid debut, short-tube reflectors based on the Newtonian reflector that feature a built-in Barlow lens.[6] They have a cheap, "fast" spherical mirror (F/ D = 4–5), leading to noticeable flaws, somewhat attenuated by a very ordinary Barlow lens increasing the focal length and reducing the image field. The basic idea is valuable, but the resulting optical quality is often dubious.

In order to reduce the tube length while correcting the field, a Schmidt lens (inherited from purely photographic, wide-field telescopes designed by Bernhard Schmidt) or a meniscus (inherited from telescopes designed by Dmitri Dmitrievitch Maksutov) is added to a Cassegrain telescope or a variant (an example is shown in Figure 2.7). The chromatism and the reduction in bandwidth (near ultraviolet and near infrared) are very limited. Numerous pure Maksutov–

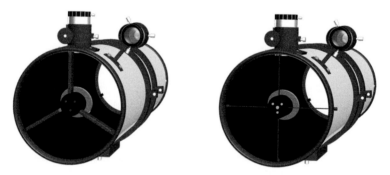

Figure 2.6 Left: in a Newtonian telescope, a thick spider brings a little more obstruction and alters the diffraction pattern; six diffraction spikes appear in stellar images. Right: the thin spider of a better-performing telescope; the vanes are of thickness 0.5 mm and four diffraction spikes appear in stellar images. A look into the tube also allows one to ensure that the tube is oversized with respect to the main mirror diameter (to facilitate the release of hot air currents) and to ascertain whether hot air can pass through the baffling (internal rings to prevent internal reflection of light), if there is any baffling.

---

[6] They are not to be confused with high-end, short-tube, Newtonian astrographs with a built-in field corrector for deep-sky imaging.

Table 2.2 Main types of pure reflectors

Pure reflectors	Notes	Particular examples (not comprehensive)
Newtonian	Primary mirror is paraboloid (close to spherical if slow F/D), secondary is planar. Typical F/D = 4–5. F/D ≥ 6 minimizes the effects of an imperfect collimation, offers a lesser obstruction, and provides a superior image for a given optical quality (wavefront error). • Obstruction, 20%–40% • Very affordable	Some models are especially designed for planets and the Moon with a "slow" F/D ratio: • 150/1200 (F/D = 8) • Discontinued/second-hand Vixen's 100/1000 (F/D = 10) • Discontinued/second-hand Vixen's 100/1000 (F/D = 10) • Discontinued/second-hand Orion Optics Ltd's 150/1650 (F/D = 11) • F/D = 6 Newtonians
Cassegrain	Primary is parabolic, secondary is hyperbolic. Typical F/D = 15. • Very few are manufactured for amateurs • Often used by professionals prior to Ritchey–Chrétien	• Astroqueyras' 620-mm[7] • Parallax Instruments, 10, 12.5, and 14 in • DFM Engineering, 16 in and wider (professional use) • Takahashi CN-212, convertible Newton/Cassegrain • Parks Optical (convertible Newton/ Cassegrain, discontinued) • Optical Guidance Systems (F/D = 8–9)
Dall–Kirkham	Primary is elliptic, secondary is spherical. Typical F/D = 12–15. Demands precise manufacturing. • Few are manufactured • Sensitive to turbulence but acclaimed quality	• Takahashi • R. F. Royce • Optical Guidance Systems
Ritchey–Chrétien	Primary and secondary are hyperbolic. Typical F/D = 8. • Demanding collimation • Obstruction 40%–45%	Widespread; optimized for deep-sky imaging with 24 × 36 format. Successfully tested for the Moon as a possible alternative to cumbersome Newtonians at the price of demanding collimation and lesser contrast.
Gregory	Primary is paraboloid, secondary is ellipsoid.	Manufactured by craftsmen only. • C. Pellier's[8] 250 mm (10 in), F/D = 31 (!) Gregory

---

[7] In France: www.astroqueyras.com/, accessible to experienced amateur astronomers.
[8] www.planetary-astronomy-and-imaging.com/en/

Table 2.2 (cont.)

Pure reflectors	Notes	Particular examples (not comprehensive)
Shiefspiegler	• Perfect for Moon but needs particular optical testing during manufacture Unobstructed, with plane secondary and tertiary. • Very delicate collimation • Excellent results if set properly	Mainly DIY. Most popular in Germany, some models are used in the USA. Kits available from www.vitastra.com/.
Herschel	Offset parabolic primary and off-axis, plane, secondary. • Unobstructed • Few realizations but excellent results	• Orion Telescopes and Binoculars's 91-mm (3.6-in) clear-aperture telescope, F/D = 13.6 (discontinued) • Discontinued Orion SVP91

Figure 2.7 A cutaway of a Schmidt–Cassegrain telescope. It has a compact and polyvalent design and the collimation is easy (secondary mirror only). Focusing is achieved by translating the primary mirror. The addition of a classic focuser eliminates the shifting effect of the mirror translation. The corrector plate acts as a support for the secondary mirror, so no diffraction spikes from spider vanes alter the image. On the other hand, the noticeable obstruction of 40% reduces the contrast while internal turbulence (as for all closed catadioptric telescopes) means that consistent performance may not be guaranteed.

Cassegrain telescopes do not need any alignment of the optics. The Maksutov variants compete with apochromatic refractors for planetary and Moon observation because they often show a reduced obstruction, and their design favors a narrow but superb image field. They are easy to manufacture. On the other hand, large-aperture Maksutovs cool very slowly and are prone to internal turbulence because of the thickness of the meniscus.

Open catadioptric telescopes (Table 2.4) adapted to lunar imaging are very interesting because, like sealed catadioptrics, they offer a large diameter and a short tube, but, with an open tube and less glass mass, they cool faster. In place of

Table 2.3 Main types of sealed catadioptric telescopes

Sealed catadioptric reflectors	Notes	Particular examples and remarks
Maksutov–Cassegrain	Meniscus and spherical primary and secondary (often an aluminized part of the internal surface meniscus). Very popular. No need for optical alignment (some models allow alignment of the primary mirror). Typical F/D = 15 with 20%–25% obstruction. Excellent results.	Widely manufactured. Because all optical surfaces are spherical, they are easily built in large quantities with very good quality. Diameters are somewhat limited because of the mass of the meniscus.
Rutten–Maksutov (Rumak)	Maksutov–Cassegrain with a separated secondary. F/D can be lowered down to 6. Excellent results.	• Discontinued/second-hand Intes/Intes- micro Alter M706, 180 mm (7 in), F/D = 6 • Santel MK91, 230 mm (9 in) at F/D = 13
Schmidt–Cassegrain (SCT)	Schmidt corrector plate, spherical primary and secondary. Typical F/D = 10–12. Average contrast because of a mean 38%–40% obstruction. Very good results if the temperature is stable; the plate is thinner than the Maksutov meniscus, hence thermal equilibrium is more quickly reached.	Widely manufactured. Modern variants have an internal field corrector for wide-field, deep-sky imaging (Celestron's EdgeHD, Meade's ACF). Some variants had an F/D = 6.3 with built-in corrector–reducer. 10–16 in. Exploited by the best planetary astrophotographers because of their fast collimation and excellent portability.
Schmidt–Newtonian	Newtonian with Schmidt corrector plate. Typical F/D = 4 with 38% obstruction.	Intended for deep-sky imaging, 50% less coma than classic Newtonian.
Maksutov–Newtonian	Newtonian with Maksutov meniscus. F/D = 4–8. Superb results if the temperature is stable.	75% less coma than classic Newtonian. • Discontinued/second-hand/ stock Intes-micro MN68, 152/ 1216 • Skywatcher MN 190/1000

the full-aperture plate corrector, they have a sub-aperture Mangin mirror: a meniscus with a reflective side. The light passes through the meniscus, is reflected on the aluminized side of the meniscus, then passes once again through the meniscus.

Table 2.4 Main types of open catadioptric telescopes

Open catadioptric reflectors	Notes	Particular examples and remarks
Hyperbolic Newtonian	Hyperbolic primary, elliptic secondary and in-built field corrector close to focal plane. F/D = 2.8–3.3	Intended for deep-sky imaging with very wide field, not suited to lunar imaging (e.g. Takahashi's Epsilon series).
Hyperbolic Baker Ritchey–Chrétien	Hyperbolic primary and secondary with two sub-aperture lenses close to focal plane.	For wide-field, deep-sky imaging (e.g. Takahashi BRC-250).
Klevtsov–Cassegrain, Mangin–Cassegrain	Spherical primary, the secondary is a Mangin mirror/corrector lens close to the primary's focal plane. Typical F/D = 8–11. Excellent results if the temperature is stable.	• TAL Telescope's TAL-250 K, TAL-200 K • Vixen's VMC 260 L, 110 L and others • Other variants with built-in train of relay lenses, like Paramythioti F/D = 6 Clavius telescopes

The conclusion is that, in the author's opinion, there is no definitive, optimal reflector for lunar imaging because each optical design is more or less adapted to an imaging site. If the temperature is steady, a pure Maksutov–Cassegrain (with no decollimatable secondary) is a very good choice, but the diameter is limited. An SCT-family optical tube is perfect for a larger diameter if the temperature is stable, and the collimation is quick and easy. A Newtonian optical tube performs very well for a limited expense and the diameter is not limited, but it requires a sturdy mount and a precise collimation of both mirrors. Other optical arrangements are better suited to wide-field, deep-sky imaging.

2.8 Mounts, tripods, lunar tracking, and exposure limit

2.8.1 Lunar imaging with a tripod; motion blurring

The exposure duration of a powerful telelens or unmotorized telescope is limited because of the effect of the Earth's rotation, the diurnal motion (Figure 2.8), resulting in motion blurring. The motion cannot be manually compensated for, even if the vertical component of the diurnal motion can be neglected when the Moon crosses the meridian (we can neglect the proper motion of the Moon). The simple horizontal component affects the image in less than a second, as we can see when we image the totality phase of a lunar eclipse.

The maximal exposure depends both on the focal length and on the size of the photosites. Tables 2.5 and 2.6 show the maximal durations for two common kinds

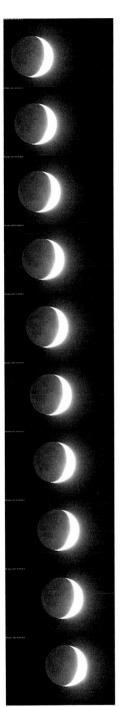

Figure 2.8 Left: a sample of ten images cropped from a movie. Relative horizontal positions are unchanged, images are separated by 12-s intervals. The instrument is a 180-mm telelens fitted to a planetary camera on a photo tripod. The westward (in terrestrial coordinates, northern hemisphere) drifting is due to the diurnal motion. This severely limits the movie duration and, subsequently, the number of images to be stacked, resulting in a poor signal/noise ratio (Section 7.3).

Table 2.5 Maximal exposure with an APS-C format sensor of about 20 megapixels (DSLR or hybrid camera). The last rows are for a telescope in place of a telelens. The longest possible exposures must be divided by two when the camera has twice as many photosites and the same sensor size. The exposures can be 1.6× longer with a 24 × 36 sensor relative to an APS-C sensor with the same number of photosites.

Focal length (mm)	Longest exposure with no motion blurring (s)	Approximate exposure setting
5	5.7	B (for bulb) exposure, 6 s
10	4.1	B exposure, 4 s
18	2.7	2.5 s
25	2.0	2 s
35	1.5	1.5 s
50	1.1	1 s
85	0.60	1/2 s
105	0.50	1/2 s
135	0.40	1/2 s
180	0.30	1/4 s
400	0.14	1/10 s
500	0.11	1/10 s
800	0.07	1/20 s
1200	0.05	1/20 s

Table 2.6 Maximal exposure with a small sensor of about 20 megapixels (compact camera)

Focal length (mm)	Longest exposure with no motion blurring (s)	Approximate exposure setting
5	2.7	"Night scene" of 2.5 s
10	1.5	"Night scene" of 1.5 s
18	0.8	1 s
25	0.6	1/2 s
35	0.4	1/2 s
50	0.3	1/4 s
85	0.2	1/4 s
105	0.1	1/10 s
135	0.11	1/10 s
180	0.08	1/10 s
300	0.05	1/20 s

of still cameras. The camera (Figure 2.9) can be screwed onto a video or photographic tripod with a standard, Kodak thread. Table 2.5 is for an APS-C DSLR (the sensor surface is about 15 mm × 22 mm, 10 megapixels, e.g. Canon EOS 350D) and Table 2.6 is for a compact still camera (the sensor surface is 1/2.3 in, about 4.6 mm × 6.2 mm, 12 megapixels, e.g. Lumix TZ10).

Figure 2.9 A planetary camera fitted to a 135-mm telelens with a T2, adjustable spacer; the camera is directly attached to a tripod with the Kodak thread at the rear if the housing. Other cameras have the Kodak thread under the housing.

2.8.2 Photo and video tripods

Tripods for still cameras have settings in three axes: pan, vertical, and tilt, the last of which is to frame in portrait mode. Tripods for camcorders generally do not offer a tilt axis, but they do have a fluid drag head for smooth movements. All have a standard Kodak screw for the corresponding thread at the base (or rear) of the camera. A powerful telelens is often heavy: the need for sturdiness of the tripod must not be neglected. Mechanical parts made of plastic and lightweight tripod legs are not reliable. Table tripods are almost useless. Serious photo and video tripods are stable and allow aiming at high altitude even if the camera is in an awkward position. At the latitude of 45° N, the Moon reaches a variable altitude: from 12° (summer) to 70° (winter). Near the equator, its maximal height is close to the zenith (90°). If the Moon is very high, we cannot comfortably see it through the viewfinder or a fixed screen, making manual focusing and exposure setting quite difficult. An orientatable screen with LiveView is highly recommended. An improvised counterpoise (such as a rock or a heavy bag) can be attached with heavy-duty adhesive tape on the base of the leg at the opposite side of the camera to stabilize the set.

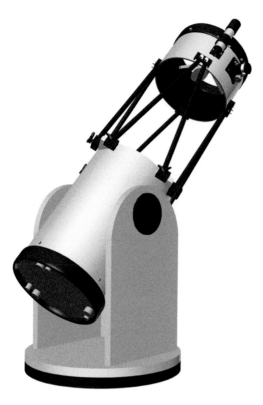

Figure 2.10 An unmotorized, Dobsonian mount. The lack of automated tracking prevents us from acquiring numerous frames while the Moon quickly passes in front of the sensor.

2.8.3 The Dobsonian mount

This unmotorized mount (Figure 2.10) does not compensate for the diurnal motion. As a result, the image quickly drifts along the field, especially with a small sensor and a strong magnification. A triple movement of the Moon appears:

- horizontal on our left hand if we are facing the Moon at meridian (in the North) in the southern hemisphere, or on our right hand if we are facing the Moon at meridian (in the South) in the northern hemisphere;
- vertical, ascending or descending, respectively before and after the meridian;
- rotational because of the inclination of the ecliptic.

The directions appear reversed in the eyepiece.

Viewed from the equator, these movements are simplified: the Moon simply rises, passes close to the zenith, and then sets, in an almost rectilinear trajectory.

This triple apparent movement is the reason why an azimuthal or Dobsonian mount is poorly adapted to astrophotography, though this does not prohibit lunar imaging. Firstly, single shots with a limited magnification may be successfully undertaken as long as the exposure is short enough. Secondly, a movie gives several individual frames, which can be re-aligned by software (PIPP, AutoStakkert! 2, AviStack, RegiStax ...) afterwards. From a movie (Figure 2.11)

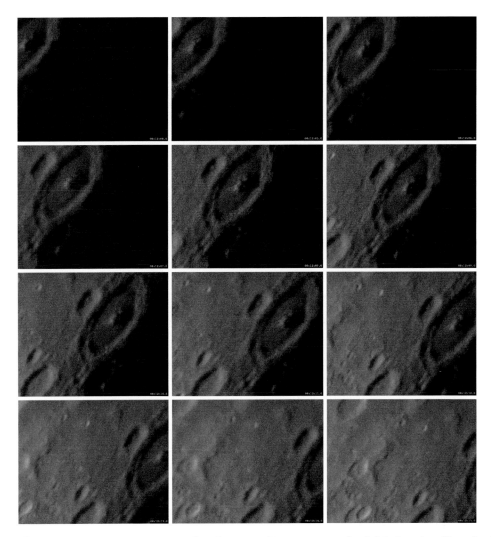

Figure 2.11 A planetary camera placed in an undriven, 460-mm (18-in) Dobsonian. Since the focal plane was located in the drawtube, the Barlow lens, albeit amplifying drifting from the diurnal motion, was necessary in order to extend the focal plane to the camera's sensor. Note the impressive, apparent motion in only 11 s. Individual frames.

taken with a 2-m-focal-length, unmotorized Dobsonian and a small, 1/3-in sensor were extracted the raw frames in Figure 2.12. Since the focal plane could not be reached by the sensor (the plane was *inside* the focuser), a 2.3× Barlow lens was added. However, the frames were re-centered and then stacked, and this provided a decent final image, at the price of severe cropping.

Numerous Dobsonians are now motorized with altazimuthal GOTO systems or equatorial platforms, and lunar imaging can be successfully performed. With an

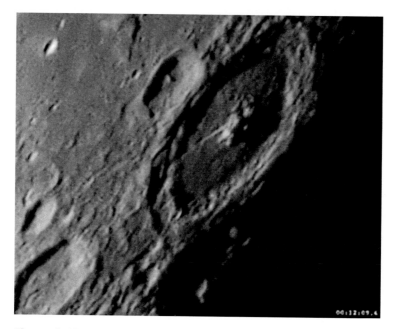

Figure 2.12 Barely fifteen frames centered on the crater overlapped during the capture. This was enough to stack and then process them. Better results may be achieved with a motorized Dobsonian or access to prime focus for the camera (60° E, 26° S).

alt–az GOTO Dobsonian, the field rotation is negligible, given the short exposures and the limited duration of the movie.

2.8.4 The German-type mount

Being pitched according to the inclination of the celestial equatorial plane, a German-type mount (see Figure 2.13) needs only one motion – the right ascension – to track the Moon, and it shows no field rotation. The right-ascension axis is generally motorized. This is a necessary condition to exploit a strong magnification. A second motor merely helps to perform delicate, "vertical" moves (in fact, declination, that is, the axis perpendicular to the celestial equator). One side maintains a counterweight; the other side maintains the optics. Unlike with most other types of mount intended for use by amateurs, a large amount of lateral room is left free for cumbersome equipment, it is not hard to aim at the zenith, and no configuration is awkward.

In addition, the attachment of any optical-tube assembly or camera fitted with a Kodak thread is easily feasible with the help of a dovetail. A sturdy, metallic ball head can be screwed though a hole. If the dovetail lacks a free hole, since it is made of aluminum, drilling a new hole is easy: the diameter is 1/4 in (6.35 mm). Dovetails exist in only two standards for amateur equipment: Losmandy and

Figure 2.13 A catadioptric optical tube on an equatorial mount. The horizontal axis is tilted to follow the celestial equator, so that movement along only one axis (manual or with a motor) is necessary in order to keep the celestial object in the field.

Vixen. The latter is for light optical tubes, e.g. up to a 12-kg load including the camera and accessories. The Losmandy standard is intended for heavier optical tubes. Many intermediate and large mounts accept both, either directly or with the help of a dovetail adapter.

2.8.5 Alt–az and equatorial fork mounts

Not all telescopes are designed for imaging, and a lightweight, compact mount may be beneficial for casual observation. Fork mounts are well suited to this purpose, and they largely equip compact, catadioptric telescopes (Figure 2.14). When we observe close to the zenith, a diagonal may help prevent a stiff neck; it can even be exploited with a camera, at the price of a little loss in brightness (only

Figure 2.14 An altazimuthal fork mount with a catadioptric optical tube. This handy, compact design is not very well adapted to lunar imaging because a cumbersome camera will generally trip over the base when the Moon is high. If a diagonal is put in place to manage room for the camera but the back-focus is insufficient, focusing may be impossible even with the help of a Barlow lens. An equatorial fork mount, or the addition of an equatorial wedge to an altazimuthal fork mount, allows imaging in all positions, like with German-type mounts.

1% with highly reflective, "dielectric" diagonal mirrors) and image quality. Sometimes, the telescope equipped with the camera cannot focus because the focal plane stays inside the diagonal (it is intended primarily for observation, and an eyepiece normally reaches the focal plane). If we remove the diagonal, the camera (especially a DSLR) trips over the base of the fork. Some fork mounts can swing vertically to occupy room left free behind the telescope, but this can lead to an awkward configuration, and should be compensated for by attaching a counterweight to the opposite leg of the tripod. However, this is not an ideal solution if the optical tube is heavy. A better solution is adding an equatorial wedge (e.g. Meade, Celestron, BC&F, Optec . . .). The altazimuthal fork mount then becomes an equatorial fork mount, perfectly suited to imaging.

One drawback of an altazimuthal fork mount is the field rotation. This is not a problem in lunar imaging because the exposure duration is brief and the rotation is compensated for by the automated registration performed by stacking software. Another drawback is that the sturdiest forks have two arms, preventing one from replacing the optical tube by another afterwards. Single-arm fork mounts sometimes offer a dovetail adapter: this solution is very versatile and allows one to put a modest photolens with a camera in place of a telescope, or to change the telescope, assuming that the load capacity is not exceeded.

2.8.6 Power supply

Current astrophotographical equipment consumes a lot of electricity. This is not a problem in a backyard or on a balcony, but imaging in the countryside demands a serious amount of electricity. Table 2.7 shows some classic examples of power consumption.

To supply the mount, and the heating resistor, if one is used, two kinds of rechargeable batteries perform well.

- Gel-sealed batteries are secured, compact, and powerful; their weight depends on the capacity. A 12-V 7Ah gel battery operates for one or two entire nights with a GOTO mount and a heating resistor, as long as slewing is not excessively utilized. Such batteries perfectly match the constant power consumption of mounts in tracking mode. The lifetime is about five years.
- Lead–acid batteries (automobile batteries) may be exploited to supply a laptop with an intermediary 12 V DC-to-220 V AC (or 12 V-to-110 V) converter. The free-air connectors must be carefully secured with insulated covers, and the batteries must not be upended. Since they contain liquid sulfuric acid, modern automobile batteries are sealed, but their use remains dangerous. In addition, they are not designed for constant use and their lifetime is shortened under such conditions.

Table 2.7 Possible power supplies required for the equipment

Mount	12 V DC, 0.5 A (small mount) to 2.5 A+ (massive, GOTO mount in slewing mode)	Mandatory for close-ups and lunar eclipses
Dew heater	12 V DC, 1 A	Optional
Laptop with old, weak internal battery	110/220 V AC, 800 mA+	Mandatory for webcam and planetary camera. Depends on laptop autonomy; needs a 12 V DC-to-110/220 V AC converter
DSLR external power supply	Depends on DSLR model	Optional, used only in the case of there being no internal battery

- Lithium-ion–polymer (LiPo) batteries are limited in charge accumulation, but they are secured and very small.

All of them require a specific battery recharger. Nevertheless, a gel battery can be charged with a standard power supply with a maximum throughtput of 10% of the capacity of the battery for about 15 hours. It is wise never to use your car battery; it will probably be flat at the end of a productive imaging session.

A last precaution, prior to traveling to a foreign country, is to purchase a plug–socket adapter. Even if most equipment is 100/110/220-V- and 50/60-Hz-compatible with the help of the secured selection switch, many sockets are not. Plug–socket adapters are widely available (for the USA/Canada/Thailand/Japan, UK, China/New Zealand/Australia, Switzerland, South Africa/Europe ...) and "universal" adapters are compatible with power supplies in numerous countries. They should be purchased in one's country of origin to match one's equipment. If the voltage is not compatible, a 100/110/220-V converter or tranformer is necessary. The sole incompatibility is for vintage mounts with an AC right-ascension motor that relies on the power-supply frequency.

2.9 Image sensors

2.9.1 What are CMOS and CCD sensors?

These two families of sensors are based on the same principle: accumulating electric charges proportionally to the incident light, thanks to the photoelectric effect. Their efficiencies can be as high as 80% nowadays.[9] Then the charges are converted into binary digits. To built an image, a sensor uses a great number of light-sensitive points – photosites – organized as a grid. A sensor is shown in Figure 2.15.

Most general-consumer, video-surveillance, and industrial image sensors, and some scientific image sensors, are CMOS (complementary metal–oxide–semiconductor) sensors, while CCD (charge-coupled device) sensors are better suited to scientific purposes such as X-ray imaging at the dentist, deep-sky imaging, spectroscopy, and photometry with professional telescopes, and some professional DSLRs.

CMOS sensors are built on standard integrated-circuit production lines, whereas the manufacture of CCD sensors demands specialized production lines, leading to a higher price. Nonetheless, because particular techniques are required for the fabrication of high-end CMOS sensors, their cost is increasing and the price is no longer a criterion when it comes to choosing between CCD sensors and high-quality CMOS sensors. The second – and main – difference is that a CMOS sensor possesses "active pixels," or, in other words, a CMOS sensor embeds numerous

[9] The efficiency can exceed 90% with some professional EM-CCD cameras; in other words, more than 90 electrons (or charges) are accumulated when 100 photons strike the sensor. This efficiency corresponds to a maximum performance, within a precise bandwidth (interval of colors). It classically drops to 40% for the deep red, and to 5%–10% for the near infrared and near ultraviolet.

Figure 2.15 A planetary camera with a USB interface. The sensor is the little rectangle at the center. Its diagonal is a little more than 6 mm (1/3 in), a classic size encountered in compact cameras and camcorders. Bridge and DSLR cameras have larger, though less high-performance, sensors (a comparison is illustrated in Figure 2.22).

functionalities[10] in its tiniest element – the photosite[11] – and other functionalities in columns of photosites,[12] whereas a CCD sensor groups these functionalities in common, shared circuits, leading to a bottleneck effect. The differences lead to a higher speed for CMOS sensors at the price of a certain dispersion of character-istics between photosites, while CCD sensors are slower but more homogeneous. CMOS sensors may address individual photosites: this is useful when one wants to acquire images only on a specified part of the sensor ("windowing" or a "region of interest," ROI), for instance for focusing at a high frame rate. Fortunately, numer-ous CCD sensors now provide an equivalent selection of photosites by skipping rows; however, the frame rate remains less than that in CMOS sensors. In addition, CMOS circuits have low power consumption and are very tolerant of voltage, while CCD sensors may need up to five different and accurate voltages, with higher current needs. That is why most portable devices nowadays are equipped with CMOS sensors.

Deep-sky cameras offer 4.5–22-μm-large photosites, while planetary cameras use 2.5–5.6-μm-large photosites. The surface area is important in order to gather more light, as if we were using a larger funnel to collect rain, but the reservoir capacity also matters because of the need to accumulate more charges. As we can see in Figure 2.16, the potential well (the reservoir) is at the bottom of the sensor.

---

[10] Charge-to-electron conversion, amplification, photosite selection.

[11] A photosite is a light-sensitive element on a sensor; a pixel (picture element) is a point on the final image.

[12] Analog-to-digital conversion, noise reduction, and FPN subtraction (e.g. Sony's Exmor® sensors).

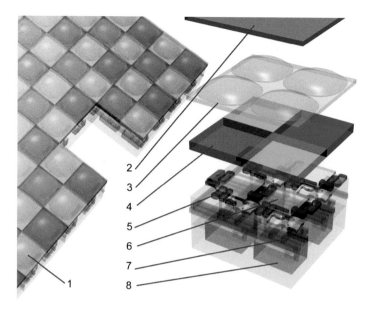

Figure 2.16 A drawing of a color, front-side-illuminated CMOS sensor, with a closer look at a quadruplet of photosites. (1) The sensor surface, viewed from above, with color filters covering photosites. Each photosite is some microns wide and can be seen only with a scanning electron microscope. There are several thousands to millions of photosites. (2) Infrared-rejection window (generally mounted at the front side of the camera and not a part of the sensor). (3) Microlenses to concentrate light. (4) Bayer matrix with one red filter, two green filters, and one blue filter (variants exist). (5) Surface components (transistors, capacitor, connectors). (6) Photosite surface (to receive incident light). (7) Underlying wiring (three or more metal layers). (8) Potential well (a kind of bucket to accumulate electric charges triggered by light incident on the photosite). In back-side-illuminated sensors, layers 5–8 are upside down; hence light directly strikes the wells without encountering any obstacles formed by surface components (layer 5) and wiring (layer 7). CCD sensors have roughly the same structure, but the surface components are less numerous and charge-to-digital output conversion is shared and located elsewhere on the chip: there are no noticeable variations in sensitivity between columns of photosites and individual photosites, unlike for CMOS sensors. Some stacked sensors have a distinct layer to separate many components from the area photosites.

Light has to travel through a certain thickness, and there are numerous obstacles: surface components and in-depth wiring.

In modern sensors, the light passes through the lens and the color filter, and then directly accumulates in the wells. The components and the circuitry are below the wells, increasing the size of the gathering surface relative to the overall surface, including obstructing surface components (the fill factor). The gain in efficiency is reported to be 30%–100%; hence the surface of the photosite is no longer an indication of efficiency with respect to classic sensors. From now on, the classic sensors will be called *front-side illuminated sensors* (FSIs); the new sensors

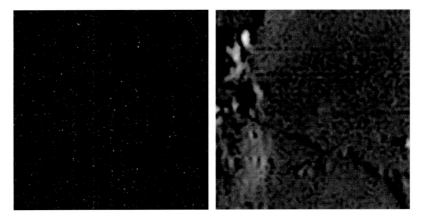

Figure 2.17 Left: a cropped and brightened image of the fixed-pattern noise (FPN) – the vertical bands – generated by a CMOS sensor (with the lens covered). The FPN is superimposed onto the frames during acquisition. The overall noise consists of random parasitic current, thermal illumination of a corner, white spots ("hot pixels") at high ambient temperature and long exposure, and differences in sensitivity (offset) between the individual photosites and between the columns of photosites. A consumer still camera automatically subtracts offset and noise; a planetary camera can neglect them except for very short exposures of the full Moon or very long exposures of eclipses. Right: horizontal banding (cropped and strongly magnified) arises from electromagnetic disturbances, low-quality cables, and electronics of insufficiently high quality. Nonetheless, the camera is an autoguider, not intended for imaging; new imaging versions with greatly improved electronics have been released.

will be called *back-side-illuminated sensors* (BSIs). The latter expression is deceptive: it does not mean that the sensor is upside down; rather, it means that the manufacturing process needs to reverse the silicon substratum (the wafer), prior to applying color filters and lenses. The datasheet indicates better dynamics and lower noise, but for now no spectacular improvement in sensitivity. For instance, Aptina's FSI MT9M034 has nearly the same sensitivity as Sony's BSI IMX174 (74% peak), but the latter exhibits lesser noise, there being hardly any fixed-pattern noise (FPN) – see Figure 2.17, with a greater margin for high-dynamic-range images (the prices are different, too). On the other hand, a recent (as of 2016) FSI sensor such as the IMX224, based on a proven technology, shows extremely little noise and no FPN, the same sensitivity in colors as a monochrome MT9M034, and a bandpass depletion extended to the near infrared.

The differences in behavior are summarized in Table 2.8.

This gives us a first opportunity to establish our needs for an ideal sensor for the Moon.

- A high sensitivity to shorten exposures, to reduce the destructive effects of turbulence.

Table 2.8 Comparison of CMOS and CCD sensors

CMOS	CCD	Consequence for lunar imaging
Each photosite has its own charge-to-voltage conversion and amplification circuits.	Photosites are read without dispersion in conversion and amplification because a common stage is used for all of the photosites after readout.	CMOS sensors show variations in response for each photosite (offset). Stacking and drizzling fix the dispersion. CCD sensors show no dispersion between photosites, at the price of a slower image readout.
For numerous sensors: in addition, each column of photosites has its own analog-to-digital converter.	A unique analog-to-digital conversion ensures homogeneity.	CMOS sensors show vertical banding due to dispersion in conversion for each column (participating in the "fixed-pattern noise"). Stacking and drizzling fix the dispersion. CCD sensors show no dispersion between columns of photosites at the price of a slower image readout.
Each photosite is partially occupied by several surface components (the surface obstruction is expressed in percentage as the "fill factor").	A small amount of the surface is occupied by additional components because "passive" CCD photosites are simpler (at the price of more complex electronics on another part of the chip).	No noticeable consequence because the Moon is bright enough for all sensors.
Low to high price (cheap webcam at $25, very high-performance astro-cameras at $380, professional cameras starting at $1000).	Moderate to high price (starting from $800).	Lunar imaging is feasible with very affordable CMOS sensors.
Very noisy signal with long exposures.	Clean signal even with long exposures (but for some extremely efficient CCD sensors especially designed for short exposures only[13]).	None, because there is no need for "long" exposures (>200 ms) except for lunar total eclipses and earthshine. Stacking reduces noise.
Rolling shutter causes image distortion. Some CMOS sensors offer a global shutter.	Global shutter.	No prohibitive effect of image distortion has been reported, because the Moon remains unchanged during shooting, even when recording hundreds of frames, hence

[13] E.g. Sony's specialized ICX618 sensor.

Table 2.8 (cont.)

CMOS	CCD	Consequence for lunar imaging
		effects of a rolling shutter combined with turbulence can be fixed by setting registration points prior to stacking.
Generally 8–12-bit converters.[14]	12–16-bit converters.	Little or no consequence if frames are stacked.
Random access to photosites plus parallel conversion of columns allows very high frame-per-second (FPS) rate (sometimes > 570 FPS).	Bottleneck because shared electronics (readout, conversion) limits frame-per-second rate (e.g. 15 FPS).	CMOS sensors may acquire numerous frames during intervals of weaker turbulence, resulting in sharper images even with noticeable mean turbulence.
Antiblooming to limit spreading over adjacent photosites in case of over-saturation.	With or without antiblooming, depending on the sensor model.	Antiblooming is perfect for lunar imaging (the drawback is the impossibility of performing light measurements for scientific purposes).
High-dynamic-range/wide-dynamic-range (HDR/WDR) capabilities (may be implemented but unexploited). Limited well capacity (the ability to accumulate charges before saturation).	No HDR/WDR capability. Larger potential-well capacity offers much greater dynamics.	No appreciable difference between CMOS and CCD sensors for lunar imaging as long as CMOS HDR/WDR remains mostly unexploited. Saturation avoidance limits clipping.
Small photosites. The loss in light gathering is compensated for with the help of new features (see the text).	Larger photosites offer greater light-gathering capacity.	Modern CMOS sensors are as sensitive as CCD sensors (for short exposures). The effective focal length has to be adapted for the correct sampling rate.

- A high frame rate to acquire as many frames as possible when the turbulence decreases – the frame rate depends both on the architecture of the sensor, especially the readout process (CMOS sensors are faster), and on the ability to acquire sub-frames to achieve very high speeds with a limited field of view (almost all CMOS and some CCD sensors offer an ROI) – and a high sensitivity to obtain short, and hence numerous, exposures in a short amount of time.

---

[14] Sony's IMX178 has a 14-bit converter.

- Low noise (few hot pixels, thermal noise, readout noise).
- A low dispersion in sensitivity between the photosites (offset) and columns of photosites (globally the FPN).
- Ideally, sensitive but tiny photosites (a contradictory wish) to increase magnification without increasing the resulting focal length. This point has to be considered carefully by the owners of catadioptric telescopes, because such optics already have a high focal length. For instance, a small Maksutov–Cassegrain might not need a Barlow lens to enlarge the image cast onto a sensor with very tiny photosites.
- A high dynamic range (potential-well capacity/converter resolution/saturation signal) to acquire simultaneously bright crater rims and dark areas near the terminator, with neither overexposure nor noise at low levels.
- In addition: a low cost . . .

Another feature may be interesting: the possibility of grouping photosites, or "binning." Binning is extremely useful for centering faint, deep-sky objects, and may also be useful with long-focus, catadioptric telescopes. A 2×2 binning, that is, grouping four adjacent photosites, amounts to virtually multiplying the photosite surface by four, reducing the magnification by four while increasing the sensitivity by four (some sensors offer an additional option of 3×3 binning). The field of view, which depends on the sensor size, remains unchanged. This is a convenient way to increase sensitivity and avoid having too strong a magnification from a long focal length, as long as the number of grouped photosites remains sufficient to provide a good image resolution.

2.9.2 Readout modes

Once the charges have been accumulated, the electronics aboard the sensor reads them. This step has a certain amount of influence on the final image, and several strategies are implemented. They even condition the scope of the sensor.

- An "interlaced" sensor alternately reads odd and even lines of the image (Figure 2.18). When such a sensor records quickly evolving images, like images of a fast motorcycle or the magnified Moon through turbulence, the still frames are striated; this is the "venetian blind effect." Stacking and drizzling can partially fix it. This technology is deprecated – but still in use for some sensors – and was designed for historical reasons, due to the inheritance from traditional broadcast equipment.
- An "interline" sensor is designed to avoid overexposure. Intermediate lines are placed between the lines of photosites to drain away excess charges. This is called an "antiblooming gate" (ABG): the purpose is to clip over-saturating charges and eliminate their propagation to prevent them from causing bright parasitic columns.
- A "progressive-scan" sensor can individually address photosites (this applies to most CMOS sensors and a number of CCD sensors). The advantage is that a small

Figure 2.18 Effect of interlacing in a camcorder with a powerful zoom and a shaky tripod, horizontally hit. The image is zoomed in. The sensor quickly reads half the frame (the first horizontal line, third line, fifth line, and so on) while the vibrations shift the image. Then alternate lines are read: the second, fourth ... The global image suffers profoundly from the time interval between the two half-frame readings. The effect is amplified with high magnifications; we conducted the same test with an industrial camera equipped with an interlaced sensor and a telescope with a focal length of 800 mm, with comparable results due to sheer atmospheric turbulence. To avoid this, we have to know whether the sensor is interlaced prior to purchasing a camcorder or video camera. "Interlaced" sensors are not to be confused with "interline" sensors; the latter are not prone to such behavior.

area only can be read in order to dramatically speed up the readout operation, with a high number of frames per second, and a reduced file size. This is extremely interesting while shooting a planet, but not very useful for the Moon.

- A "full-frame-transfer"[15] sensor (such sensors are mostly CCD sensors) is divided into two parts: the first part gathers incoming light, then charges are transferred to the second part prior to conversion and reading. Such a sensor is extremely sensitive but can read only entire images, at a slow rate. It often has no antiblooming feature. A full-frame-transfer sensor is perfect for deep-sky imaging and photometry, but not for lunar imaging, which requires a high frames-per-second rate.

2.9.3 Sensor cooling for short exposures?

To avoid hot pixels and a certain background gradient due to heat, professional deep-sky cameras are cooled with a flow of nitrogen, whereas amateur cameras use a thermoelectric Peltier cooler (like a camping cooler). This has little influence in lunar imaging, but knowing that is useful prior to purchasing the camera if we are planning to carry out long exposures for lunar eclipses. However, some planetary cameras do have a Peltier cooler to limit internal heat (e.g. QHYCCD's IMG0-series cameras). Other CMOS-sensor cameras are cooled to limit the noise during long exposures (e.g. I-Nova's Plb-Mx2 Nebula and ZWO's cooled ASI185MC, among

---

[15] This concerns the readout process, not the "full-frame," 24 × 36-mm image format as in a professional DSLR.

Figure 2.19 A single exposure of a lunar eclipse taken with a CMOS DSLR. The lightness has deliberately been exaggerated to reveal the background noise. Left: a "hot pixel" corresponding to a blue photosite with a very low threshold level. It cannot be confused with a real star (right) resembling a disk because its light has been spread during the 5-s exposure, while the hot pixel remains circumscribed within a single photosite. As CMOS photosites always show a variation in sensitivity and minimal signal level (offset), the background seems unevenly colored.

others). This issue concerns DSLRs, too (Figure 2.19). Some astrophotographers use a cooled housing for their DSLR in order to maintain noise at low levels. Dry ice was a solution (cryocameras), but modern cooling is performed by a thermoeloectric Peltier device in a sealed cabinet (e.g. extruded polystyrene). Fortunately, the longest exposures during imaging of lunar eclipses last only 5–10 s: the noise remains limited, especially with high-end CMOS sensors and CCD sensors.[16] Subtracting one, or preferably several, "darks" (images taken with the shutter closed, at the same ambient temperature and with the same exposure duration) with the help of software (e.g. IRIS, DeepSkyStacker . . .) is an efficient and very easy way to wipe out hot pixels. This technique has been inherited from deep-sky imaging, and subtraction can even be performed in real time with FireCapture, for instance.

2.9.4 Global vs. rolling shutters

After the charges have been accumulated for the desired exposure time, the sensor has to close: this is the role of the shutter. Modern shutters are purely electronic, with no vibration, and two different kinds of them exist.

- A rolling shutter closes the photosites sequentially. The sensor reads the first photosites before completion of the accumulation of charges in the last photosites. In other words, an image taken with a rolling shutter is not a snapshot but a continuous process. When the sensor captures a quickly evolving image (a fast motion), the resulting image seems distorted. One might fear that there would be unfixable distortions in images taken at high magnification with turbulence. Fortunately, when we stack images, tens of "registration points" allow the software to re-align recognizable parts of the image.

[16] With the exception of some specialized sensors like the ICX618, intended for short exposures at high FPS with high sensitivity.

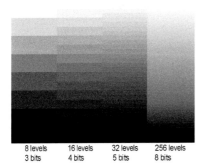

Figure 2.20 The converter resolution determines the accuracy of the translation of shades into quantized values.

8 levels   16 levels   32 levels   256 levels
3 bits   4 bits   5 bits   8 bits

- A global shutter instantaneously prevents all photosites from receiving light. The image is a real snapshot. This feature is offerered by a number of CMOS sensors and most CCD sensors.

These two technologies have proven to be equally well suited to lunar imaging.

2.9.5 Converter resolution

After the shutter has closed, electric charges accumulated in the potential wells have to be converted into voltages (the analog signal) and then into bits, or binary digits (the digital signal), that the computer can understand. This is done by the analog-to-digital converter (ADC). An 8-bit converter can translate an analog voltage into up to $2^8 = 256$ quantized levels, or, in other words, numbers between 0 and 255 (Figure 2.20). This does not mean that the dynamic range is 256, because the difference between high and low levels of light was already limited by the potential wells' capacity during acquisition. The range of levels is the converter's *resolution* (not to be confused with optical resolution). This quantization is the second step in reducing the dynamic of incoming light to a final range of values. A 16-bit converter can translate up to 65 536 levels (per color). Most DSLRs have a converter resolution of 12–14 bits.[17] Deep-sky cameras have 16-bit converters. Planetary cameras have mostly 8–12-bit converters, sometimes 14 bits. Once again, stacking helps to produce 16-bit or 32-bit images, statistically recovering a higher dynamic.

Therefore, we may be facing two kinds of situation.

- Low magnification (earthshine, conjunctions . . .). The dynamic range is maximal. This requires a high-dynamic-range sensor and a 12–16-bit converter. If this is not the case, we have to rely on a workaround by combining two or more exposures – by preference short movies to be stacked – taken during very limited time intervals, for instance five or ten seconds, for a quickly evolving phenomenon a conjunction or an occultation.

---

[17] They are underexploited when images are stored in 8-bit, JPEG compressed format.

- High magnification (lunar close-ups). This requires an ordinary dynamic range from the sensor, with an 8-bit converter. As shooting conditions differ widely according to the lunar phase and the targeted region, we may only evaluate the dynamic range during several tries, and then adapt the magnification to the camera's capability. In doing so, a variable-length tele-extender and/or a set of eyepieces with various focal lengths, or a set of spacers behind a non-telecentric Barlow lens, can help us to reach the right magnification, and hence to lower the dynamic to match the camera's capabilities.

2.9.6 High-dynamic-range (HDR) sensors

Some subjects exhibit extremely high contrast: conjunctions, occultations, earth-shine . . . The difference between highest and lowest light levels may greatly exceed the capacity of the sensor, leading to overexposure while low levels are correctly exposed, or, inversely, underexposure while high levels are correctly exposed. This is directly related to the full-well capacity, that is, the ability to acquire and store electric charges before over-saturation occurs, like water flowing from a full bucket. If the capacity is limited, faint light levels are not recorded while high levels already reach the saturation level. The difference between high and low levels is *dynamic* (expressed in decibels, or dB); it is related to the potential wells' capacity: the size of the silicon substrate of which the photosite's accumulated-charge reservoir is made. Deep-sky imaging favors large photosites, e.g. 10–22 µm in width, and deep potential wells, e.g. 60 000 electrons in capacity, roughly 3–5 times better than for planetary cameras. Unfortunately, for reasons relating to cost and to accompany the reduction in size of portable devices, manufacturers reduce the sensor size, and many planetary cameras have 2.5–5.6-µm photosites and a well capacity of 10 000–16 000 electrons.[18]

Manufacturers compensate for this drawback by adopting new features such as double-sensitivity sensors. For instance, Fuji S-pro series DSLRs use, on the same sensor, both tiny photosites for high levels of light without saturation, at the price of having shadows that are too dark, and large, sensitive photosites for low levels of light, at the price of saturation of the brightest areas (an SR sensor: S are large photosites and R are tiny photosites). The DSLR internally combines signals from small and large photosites, and the result has a high dynamic range. In other words, such cameras correctly shoot scenes comprising simultaneously low light levels (e.g. shadows in backlight) and high light levels (e.g. sunlight on white clothes, sand, or snow). For example, if the highest light level is 1000 times the lowest light level in the original image, the image dynamics is $10 \times \log(1000) = 30$ dB. With a difference of, say, 12 magnitudes between a lunar crescent (high light) and the part of the Moon which is illuminated by earthshine (low light), the dynamics is $2.512^{12} = 63\ 129$, or 48 dB. So we must choose a sensor with a dynamic range of at least 48 dB. This value is – sometimes – indicated in the sensor's specifications document (datasheet) by

---

[18] The IMX185 has a capacity of 22 000 electrons.

the manufacturer, or in the specifications of the camera. For instance, a Sony Exmor® IMX104-based, Backfly PGE-14S2 C color camera has a dynamic range of more than 73 dB, greatly exceeding our average needs (with the reservation that, because the signal/noise ratio, although very good, is 43 dB, stacking is required in order to lower the noise, as with numerous cameras).

Another solution is the built-in HDR[19] capability. Some sensors are able to internally combine a short exposure and a longer exposure; this mimics a dual-sensitivity sensor. For instance, the Aptina MT9M034, fitting several cameras from various manufacturers, is claimed to have a dynamic range of 115–120 dB. It takes a short exposure, stores data, takes a second slower exposure, compensates for slight drifting of the image (in the case of high-speed motion of the subject, such as with turbulence at high magnification), and then internally combines the two exposures before delivering a 12-, 14-, or 20-bit image with a 120-dB dynamic range.[20] Some cameras such as QHYCCD's QHY5 L-II implement this 120-dB range with the provided EZPlanetary image-acquisition software. The technique known as digital overlap has the same purpose, but in this case each line is read with three successive different exposures, resulting in an HDR, three-exposure frame. As an example, the Sony IMX224 sensor provides an HDR capability, and it achieves a 70-dB dynamic range. The function requires a specific image signal processor.

HDR/WDR image brightness rendition is intentionally non-linear, for example it is logarithmic, or programmable with several "knees" or inflexion points. Then the output image is transcoded into a more limited range, like tone mapping (Section 8.12). Since HDR/ WDR techniques were developed for video surveillance in high-dynamic-range scenes, these functionalities are suited to high-contrast events such as earthshine and conjunctions. Some functions may not be implemented when cameras and their associated software are available, or the functions are implemented later in updated drivers, or they require an optional circuit. There is a vast potential gain for lunar imaging at low magnification and very high contrast in future HDR/WDR implementations. However, manual HDR imaging is feasible with the help of specialized software or with classic image editors.

2.9.7 Color sensors vs. monochrome sensors

Most sensors are actually monochrome only: they do not distinguish colors. Like the human eye, they need several specialized receivers, some for the red, others for the blue, and so on. According to Helmholtz, whose theories were based on Young's work, to perceive the colors of a rainbow one needs a three-bandwidth system: blue, red, and green. To mimic the response curve of the human eye, the sensitivity to

---

[19] Also called WDR (wide dynamic range), possibly with various technical solutions.

[20] According to the datasheet – this may have changed in subsequent versions – this option is available only in a particular HDR setting of the sensor, using an internal, serial, "HiSPi" communication bus, requiring the image-acquisition software to take into account this functionality.

green has to be twice the sensitivity to the other two colors. The simplest way to register colors is by placing a matrix of colored filters in front of a monochrome sensor. Each colored filter corresponds to a photosite. The obvious consequence is that a sensor intended to provide one-megapixel images needs one million photo-sites for red, another million for blue, and two more millions for green. Since most sensors actually use twice as many "green" photosites to retrieve the color balance of the human eye, a color sensor must have four times as many photosites as the resulting number of color points in the final image. The set of color filters in front of photosites, also known as a color filter array (CFA), may form various patterns. The most common is the Bayer matrix, named after its inventor, Bryce Bayer. As mentioned previously, it comprises two green filters, one red filter, and one blue filter. Since four photosites are needed to reconstitute one color pixel, the actual image resolution of a 20-megapixel sensor is only 5 megapixels. Nonetheless, a process called debayering redistributes color information (using matrix calculus) and calculates artificial, intermediate pixels (interpolation). The apparent loss in resolution is only 45%. Raw images are not affected by debayering.

The second consequence is that the sensor has less sensitivity because of the colored filters, each of which totally dims a part of the incoming light while only the desired color passes through. The color filters themselves cause a loss in sensitivity, typically about 30% (e.g. IMX136).[21] It is undisputed that color sensors show both less sensitivity and less resolution than their monochrome equivalents. But since color sensors are in high demand for general-consumer applications such as smartphones, manufacturers are making efforts to improve performance regarding sensitivity, despite the ever-present drive to reduce the size of photo-sites. This is a real challenge.

Others matrices exist, including a monochrome (panchromatic) photosite. This additional, unfiltered photosite increases the sensitivity and bandwidth up to the near infrared and near ultraviolet, with no gap in crossover regions between color filters. Although such matrices are not very common at the time of writing, one, called Truesense Sparse Color Filter Pattern, is already being used in some sensors (e.g. KAI-02170), with a claimed gain of a factor of 2–4 in sensitivity. Color sensors with no filters are available, such as the Foveon X3®. It takes advantage of the differential absorption of colors by silicon (the basic material of sensors and other chips) depending on the depth of penetration of light of different colors. The upper layer of photosites captures blue, the second layer captures green, and the third one, at greater depth, captures red. Unlike in CFA sensors, each photosite corresponds to each detail of the image. In other words, an X3 sensor has three times the accuracy of a Bayer CFA sensor (about double the accuracy if we consider the interpolation). At present, it is in use in Sigma DSLR, Toshiba TELI scientific/industrial cameras, and other devices.

---

[21] The peak quantum efficiency of a color sensor is only about 15% less than that of its monochrome equivalent, but the overall loss can reach 45%, mainly because of the crossover gaps.

Consequently, we have five possibilities, but only the two first are common.

- Choosing a monochrome sensor is the best technical choice: better sensitivity (no filter), better resolution (each photosite ideally corresponds to a specific point of the image), better bandwidth, including the near infrared.
- Choosing a standard, more widely produced color sensor, which is cheaper, is the best budgetary choice. It is often the only possible choice with webcams, entry-level planetary cameras, or DSLRs (the "black and white" mode is simply a software cancellation of colors, but this does not remove the CFA).
- Choosing a Foveon color sensor (some DSLRs only).
- Choosing an alternative color-matrix sensor, including both photosites with RGB filters and photosites with no filter.
- Acquiring movies in raw mode (with no debayering), and then processing them as a monochrome movie.

Note that the last possibility represents the emerging technique of using "all-purpose" cameras with sensitive, near-infrared-depleted color sensors[22] as monochrome cameras if the sensors are not available in a monochrome version.

Some daring astrophotographers remove the Bayer matrix (along with the microlenses) of their DSLR to retrieve the full resolution and to increase the sensitivity in the deep red. This risky and radical operation voids the guarantee and may damage the sensor, but it is worth mentioning if one can dedicate a spare DSLR to full-resolution, monochrome or narrow-band imaging of the Moon or emission nebulae. See

http://stargazerslounge.com/topic/166334-debayering-a-dslrs-bayer-matrix/
www.reddit.com/r/astrophotography/comments/2oaoul/debayering_its_what_
ive_been_doing_during_this/

DSLRs like the Olympus OM-D E-M5 II and the Pentax K-3 II can slightly move their sensor for each photosite to pass behind each filter of the Bayer matrix. During this four-step displacement, each photosite successively acquires a color signal, hence no destructive interpolation occurs. This requires a stable target, hence a motorized mount is necessary to track the Moon while shooting.

2.9.8 Sensor format, number of pixels, and resolution

At a first glance, lunar imaging requires a large sensor, because the Moon is large, and a great number of pixels, because the Moon has numerous, tiny features.

---

[22] The author obtained full-resolution, non-interpolated monochrome images including in the deep red and near infrared with an ASI224MC color camera in beta testing; the movie was recorded with FireCapture with the "debayer" option unchecked; the movie was stacked by RegiStax with no debayering and the final image showed no CFA pattern at all, thanks to drizzling and raw images.

A sharp image is commonly called a "high-resolution" image. But we have to consider the exact meaning of resolution, which is the ability to get a sharp image of a small crater, or, more exactly, to get the best possible image of lunar details with respect to the optical performance of the instrument. This is related to the sampling rate, which is the measure of the relationship between the width of the tiniest visible lunar details and the width of the individual photosites. If the two values coincide, we get a high-resolution image. A sharp image may be small because the sensor has few photosites, for instance 300 000 photosites for a webcam or a high-end planetary camera. If we use a more sensitive sensor with 300 000 large photosites, the noise is lower, and the exposure is shorter, but we need a higher magnification in order for the size of lunar details and the width of photosites to match. The resulting resolution of the image is exactly the same! If we use a megapixel sensor (at least 1 000 000 photosites), assuming that the size of the photosites matches the size of the smallest lunar details, the image has the same resulting resolution, but the field is larger.

That is why the resolution does not rely on the number of pixels, despite the fact that vendors of consumer cameras often claim that a 20-million-photosite camera is better than a 5-million-photosite one. Moreover, for a given size of the sensor, more photosites on the same surface implies smaller photosites and wells, and hence more noise and less sensitivity. In the final analysis, we must consider clear definitions for the following terms.

- Resolution: the width of the smallest appreciable lunar details on the image.
- Sensor format: the size of the sensor (see Figures 2.21 and 2.22), conditioning the width and height of the image field, with no relationship with either the number of pixels or the resolution (see Table 2.9).
- Number of pixels: the real number of photosites in the case of a monochrome sensor. In the case of a Bayer-matrix color sensor, the effective number is lowered to 45% on average and the final image is interpolated to show as many pixels as from a monochrome sensor.

2.9.9 Sensor efficiency

Efficiency is, for a sensor, the capacity to detect faint light. An efficiency of 75% means that a quarter of the light is wasted while 75% is translated into a usable, digital image. The efficiency is lower for certain colors. When we look at the published data, the efficiency is always given for the maximal sensitivity at a given color, for instance green or blue. We will see later that some telescopes take advantage of green, while others are better in the near infrared: that is why we have to take a closer look at the efficiency for each color, even for the Moon.

An astronomical image is always altered by turbulence at different speeds. Turbulence manifests itself at two (or more) speeds at the same time, for instance

Table 2.9 This table lists the sizes of some representative and extreme sensors. The diagonal has to be considered because different telescopes have different undistorted field widths. Slow-F/D (F/D ≥ 8) telescopes and astrographs have a larger undistorted field, and perfectly match large sensors.

Standard format size (inches)	Actual dimensions (mm)			24 × 36 ratio
	Width	**Height**	**Diagonal**	
1/4-in ICX618 (e.g. PLA-Mx)	3.67	2.6	**4.50**	9.6
1/3-in MT9M034 (e.g. ASI120MM) (*)	4.83	3.63	**6.04**	7.2
1/2.5 in	5.76	4.29	**7.18**	6.0
1/2.35 in (e.g. Lumix TZ10 = 1/ 2.33 in)	5.80	4.27	**7.20**	6.0
1/2.3 in	6.16	4.62	**7.70**	5.6
1/2 in ICX267AL (e.g. Basler A631 F)	6.40	4.80	**8.00**	5.4
1/1.8 in	7.18	5.32	**8.94**	4.8
1/1.7 in	7.60	5.70	**9.50**	4.6
1/1.6 in	8.00	6.00	**10.00**	4.3
2/3-in ICX285AL (e.g. Atik 314)	8.80	6.60	**11.00**	3.9
1 in	13.20	8.80	**15.86**	2.7
4/3 in square KAI-04022 (e.g. Atik 4000)	15.15	15.15	**21.43**	2.0
4/3 in	17.30	13.00	**21.64**	2.0
1.5 in	18.70	14.00	**23.36**	1.9
APS-C Sigma DP1	20.70	13.80	**24.88**	1.7
APS-C Canon	22.30	14.90	**26.82**	1.6
APS-C Sony, Pentax, Nikon	23.70	15.70	**28.43**	1.5
APS-C Sigma SD1	24.00	16.00	**28.84**	1.5
APS-H Canon	27.90	18.60	**33.53**	1.3
24 × 36 KAI-11002 (e.g. Sbig 11000)	36.00	24.00	**43.27**	1.0
Leica S2	45.00	30.00	**54.08**	0.8
Pentax 645D	44.00	33.00	**55.00**	0.8
Leaf Aptus II-12	53.70	40.30	**67.14**	0.6
Megacam (3.6-m CFH Telescope)	300.00	300.00	**424.26**	0.1

(*) The 1/3-in format is somewhat approximated (being a commercial denomination), for instance the IMX104 is slightly larger, with a diagonal of 6.28 mm.

a slow turbulence at some tenths of a hertz (Hz) – the number of events per second – and another speed of some hundreds of Hz, depending on the different layers of the atmosphere, local and high-altitude winds, and heat emitted by roofs, tarmac, and concrete. The turbulence speed is often greater than the exposure duration. If the exposure is very short with respect to the turbulence, it is advantageous to freeze the image. If the exposure is long, the resulting image is a kind of average of the

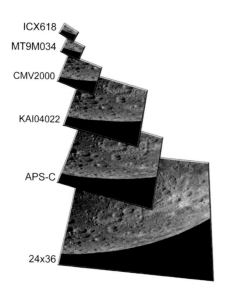

ICX618
MT9M034
CMV2000
KAI04022
APS-C
24x36

Figure 2.21 The size of the sensor determines which surface of the image cast by the optics is acquired: this is the field width. The resolution is the measure of the smallest appreciable details in the image. Increasing the number of photosites enlarges the image but does not provide a better resolution, assuming that the individual photosites have the same surface area and the magnification remains unchanged. Here is a comparison between the surfaces of some representative sensors. For instance, the MT9M034 has tiny photosites, resulting in a fine resolution for a given magnification, while the ICX618 has fewer but larger photosites, resulting in an excellent sensitivity, a larger field width, but a lower resolution if the magnification is unchanged. The 24 × 36 format applies to professional DSLRs.

Figure 2.22 A comparison of the surface of two sensors (the rectangles inside the cameras). Left: a 1/3-in MT9M034 CMOS sensor inside a cooled planetary camera. Right: an APS-C (more precisely, DX), ICX413AQ CCD sensor inside a Nikon D70 DSLR. While the DSLR has the wider field by far, perfectly suited to lunar eclipses or to frame the entire Moon, the planetary camera gives substantially better results in high-resolution lunar close-ups. Thus, the cameras have complementary usages.

turbulence; in other words it is blurred – an unevenly spread motion blurring. If the sensor has a high efficiency, this decreases the effect of the turbulence. In addition, it can record images even at a higher magnification despite the attenuated brightness.

The efficiency of a sensor can be expressed in two ways.

- The quantum efficiency is the percentage of incoming light actually gathered by the sensor. For instance, it can acquire 65% of the blue and 30% of the red. This is the most convenient way to compare sensors. Some manufacturers publish the quantum efficiency of their sensors (e.g. Aptina's datasheet for professional customers).
- The relative efficiency indicates the sensitivity in different colors with respect to the color for which the sensor is the most sensitive. This color, for instance the green, is notionally rated 100%. This does not mean that the sensor converts 100% of this color. This is done by some other manufacturers (e.g. Sony).

The datasheet specifications may show different measurements of the sensitivity, e.g. the number of electrons per ADU (analog-to-digital unit) (that is, the number of electrons after charge-to-electron conversion needed to increment the digital value after the analog-to-digital conversion), or the number of volts per lux per second, or, more commonly, the output in millivolts under given acquisition conditions (e.g. 1/30th of a second exposure with an F/5.6 lens). This does not facilitate comparisons!

The efficiency is represented by a curve (illustrated in Figure 2.23).

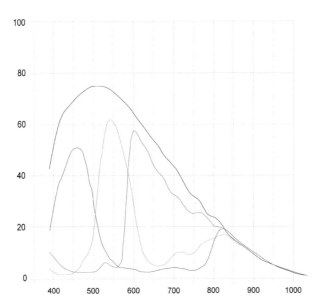

Figure 2.23 The quantum efficiency of a sensor. The black curve is for the monochrome version. The maximal efficiency (vertical scale) is obtained close to the green. The same sensor in its color version shows a lesser sensitivity in all colors (wavelengths in nm along the horizontal scale) due to the filtering effect of the Bayer matrix. The efficiency of the monochrome version is 20% better in the blue (450 nm) and 10% better in the red (650 nm). In crossover zones, the difference is amplified, especially at about 500 nm, where the difference in sensitivity is as great as 45%. Because filters exhibit leakage at large wavelengths, the two versions are equally good in the near infrared. This curve was redrawn from a datasheet of a real sensor.

- A curve with a peak in the red indicates that the sensor is suited to imaging in the red and near infrared, and this is very interesting as a means to decrease the effects of turbulence with large telescopes.
- A curve with a peak in the green is of interest in order to best exploit an achromatic refractor (such refractors are often optimized for about 550 nm).

Furthermore, the comparison between the monochrome version and the color version of the same sensor illustrates the difference in sensitivity due to the Bayer matrix.

2.9.10 The frame-rate dilemma with a planetary camera

It is often claimed in forums, astronomy clubs, and advertising, that cameras which offer a high frame rate (frames per second or FPS) are better because this feature helps to combat turbulence. It is true that turbulence may decrease for a few seconds, from time to time. At such moments, the ability of the camera to catch numerous, sharp frames is decisive. But certain preconditions have to be fulfilled:

- most importantly of all, accurate collimation and focusing prior to the start of image acquisition;
- a sensitive sensor for short exposures;
- a high-performance camera-to-computer throughput (for planetary/industrial cameras).

Basically, the frame rate is related to the inverse of the exposure time. If we set the exposure to 20 ms, the acquisition rate should be 1/20 ms = 50 FPS. This is if no bottlenecks or weaknesses exist in the acquisition chain, such as sensor shutter speed, readout and conversion speed, camera-to-computer interface throughput, acquisition-software reactivity, buffering, and hard-disk or solid-state-disk access speed.

Cameras have a speed limit. For instance, an excellent camera equipped with a renowned – albeit deprecated – IXC445 CCD sensor and a 12-bit converter delivers a relatively slow rate of 30 FPS at full resolution, even with a fast USB3 interface. In this example, a wide field of view and a high image quality prevail over the FPS rate. In this case, 150 images may be sufficient for stacking, because this sensor has a very low noise level. This requires 150 images to be acquired at 30 FPS, meaning that it takes 5 s to capture frames at the maximum FPS. At the maximum of 30 FPS, the duration of the longest exposures is limited to 1/30 s = 33 ms. Nonetheless, this camera allows fastest exposures of duration 100 μs (0.0001 s). Obviously, this does not imply that the maximum FPS is 1/0.0001 = 10 000 FPS!

Let us study another camera, with the quite different MT9M034 CMOS sensor. The camera is fast, with a roughly comparable number of photosites, the same USB3 interface, and the same 12-bit converter. The maximal claimed FPS is 254, which is very different from the previous example. But this is true only with a reduced resolution of 320 × 240 pixels and an 8-bit conversion. At full resolution, the FPS is 60, allowing the capture of twice as many frames as in the previous example, but this implies that the maximum exposure is half that in the previous example, or 1/60 s = 16 ms. If we need longer exposures, like 33 ms, the FPS is stressed by the exposure time rather than by the intrinsic capabilities of the camera and its interface. In this case, the two cameras offer strictly the same performance at full image size and with the same converter resolution. Anyway, the sensitivity and noise characteristics are different, possibly leading to different uses of the cameras. As of 2016, an IMX174-equipped camera can attain 740 FPS for 320 × 240 pixels and 397 FPS in 640 × 480 pixels.

A degree of freedom is offered by setting the gain. The gain is the signal amplification after the potential wells have accumulated charges. In other words, this is strictly equivalent to setting the ISO in a DSLR, or amplifying low light levels in any image-editing software.[23] Nonetheless, it provides a convenient way to shorten exposures when we want to acquire a large number of frames within a limited duration, especially when turbulence suddenly becomes minimal. Gain has limits, and it strongly increases noise. This can be overcome by acquiring more frames to be stacked. But acquiring more frames implies a longer acquisition duration, and the turbulence may increase in the meantime.

Now we can consider an IMX224 sensor. Its main characteristics are similar to those of the MT9M034, but, since its readout noise is extremely low, an enormous (and not very useful) 72-dB amplification is possible. The noise remains negligible up to a 30% amplification. Hence we can reduce the exposure by a factor of two and increase the gain with no noticeable additional noise, resulting in sharper images if the main speed component in turbulence is commensurate with the shutter speed. Once again, high sensitivity and low noise are more important than a high FPS, assuming that the turbulence does not evolve on a time scale of minutes.

2.9.11 Electric disturbances, computer performance

As of 2016, most planetary cameras are essentially insensitive to external perturbations and do not have excessive current needs. But, just in case, it is best to know about some classic common sources of perturbation during image acquisition. The results are bands in images, sudden loss of connection between the camera and the software, or software freezing.

Here is a list of key items to survey.

- Electromagnetic disturbances: smartphones in the vicinity, parasitic radio emissions from switching-mode power supplies of laptops or desktop screens. If we use a reel, it is important to unreel it before connecting devices. (The reel forms

---

[23] Some image-acquisition software provides both a digital gain comparable to ISO setting and an electronic gain performed by the sensor. The first one is practically useless.

Figure 2.24 This image of Jupiter was acquired while the laptop was plugged into a wound reel acting as a coil. This emitted electromagnetic disturbances, which affected data transfer via the low-quality USB cable of the webcam. After the reel had been unwound, these disturbances totally disappeared. Comparable artifacts may occur when the camera is plugged in via a hub or an extension cable, or when the laptop or the desktop screen is powered by a "noisy" switching-mode power supply.[24] Image by Maxime Giraudet (this test image is not representative of Maxime's gift for planetary imaging!).

a coil emitting an electromagnetic field. The author and a colleague have noticed such behavior, as in Figure 2.24.)

- Power supply: USB interfaces supply at best 500 mA for desktops, less for laptops or low-end desktops. If the camera does not work perfectly with a laptop, we can try using a desktop with a high-quality motherboard. The author solved recurrent freezing and image-banding problems of a camera equipped with a MT9M001 sensor, which was very sensitive to power-supply concerns, by plugging the camera into a desktop fully meeting the USB current-supply requirement.
- USB overload: generally, two USB ports are handled by one controller (current and data-stream management). It is safer to plug the camera into one USB port and leave the second port free, rather than plugging devices into both of a pair of USB ports.
- Insufficient transfer rate for the interface to the camera, e.g. when a USB3 camera is plugged into a USB2 port.
- Insufficient buffer memory. During the acquisition of a large number of frames, a small part of the volatile memory of the computer[25] is used as a buffer, because a hard disk – where frames are to be recorded – is one of the slowest

---

[24] Other comparable disturbances may occur simply on moving the mouse. Such disturbances lead to changes in the laptop's power usage when it is powered solely by its battery. Some cameras are very sensitive to power-supply variations (e.g. interrupt requests – IRQs – triggered by devices like a mouse or hard-disk drive, HDD) and radio frequencies.

[25] Managed in different locations: CPU cache, DMA, disk controller ...

parts of a computer. When the memory is full, there is a bottleneck. This may require some workaround like using a RAMDisk (Appendix 2), increasing the RAM if the acquisition software manages a memory buffer (like FireCapture), or using a fast SSD (solid-state disk).

- Using a USB extension cable or a USB hub perturbs the data stream. The camera should be connected directly to the port.
- Using a poor-quality cable. Numerous consumer USB cables are poorly shielded or manufactured with low-quality copper. A good USB cable is rather rigid, indicating that it has a strong wire section, reliable ground wire, and correct shielding.

Since the computer's main duty is to continuously transfer frames from the camera to the disk, little calculating power is needed. But lunar imaging is much more demanding for the computer than deep-sky imaging because the disk and some components[26] are massively utilized for frame transfer. A good measure to preserve performance is to temporarily close down all non-vital resident software applications such as antivirus programs, automatic updaters, screen savers, and listeners for mobile devices. The screen may be dimmed to save power. Stand-by mode also saves power.

The best thing to do is to wait for low-turbulence conditions, which often occur after midnight when the atmosphere has cooled, before turning the laptop on. A fairly new laptop may operate for four hours, but the battery's capacity soon depletes with time. Unfortunately, they spontaneously grow old even if they remain unused, and a new battery may represent a third of the price of the laptop.

As of 2016, a budget laptop, with a Core I3™ processor, 4 GB of memory, and a classic 5400-rpm hard disk performs well enough to acquire 260 FPS with USB3. We just need sufficient room on the hard disk. As a typical example, assuming that we have a monochrome camera, acquiring 500 frames of 1280 × 1024 pixels as a SER file, the file size is 625 MB. During a stable night, a good crop may represent 30 movies × 625 MB = 18 GB, equivalent to four DVDs.

Stacking movies is another challenge. Several tens of movies per night have to be stacked prior to final image processing. If we have a powerful desktop, movies can be transfered to it. This may require some tens of minutes with a USB key, depending on the number and duration of the movies (45 minutes is the mean value for the author at this time). We can also stack movies even on a laptop with a limited computing performance. This step typically takes ten hours with a good crop of several tens of movies. A more convenient solution is stacking in batch mode on the laptop, for instance with AviStack or AutoStakkert! 2. The software runs all night long, stacking the movies one by one with no human intervention,[27]

---

[26] CPU, PIO, and DMA, including the USB3 controller.

[27] AutoStakkert! 2 needs manual operation for the very first movie, but then it automatically processes the remaining movies. Multiple movies can be loaded by dragging and dropping onto the main window.

and then the stacked images can be transfered in minutes to a more powerful desktop for final processing. Another solution is to write movies on-the-fly on an external USB disk. The transfer rate is more than 500 MB per second in USB2 and SATA III, or twice that for USB3. SSDs are a very convenient solution for laptops because they contain no moving parts, and this is an asset in terms of the reliability of mobile devices. On the other hand, their limited capacity is, as of 2016, an obstacle if we want to acquire numerous large movies. However, this feature should evolve quickly. Removable SSDs offer (or will offer) a perfect solution with a limited power consumption (a conventional hard-disk drive – an HDD – needs up to ten times as much electric power), a fast transfer rate (at least the same as that of a fast HDD), increasing capacity in coming years, a superior reliability, and the possibility of serving as a shuttle.

2.10 Image and video formats

2.10.1 JPEG compressed images

JPEG (Joint Photographic Experts Group) images are often used because their use saves storage space, and, due to their limited size, they are easy to share via the Internet even with a limited-bandwith connection. This is why they are extremely common on the Web. This is due to compression. If an image contains uniform areas (e.g. a large, full black area like the background sky around the Moon), there is no real reason to repeat the same value for each point: black, black, black ... It is better to say that the next 100 points are black. This is the basic principle of compression. Moreover, our eye and storage constraints on Web servers are adapted to no more than about 256 different colors, while original astronomical images often contain 65 000 colors. Compression also affects the image's accuracy: even if the image is slightly blurred, the subject remains recognizable. The JPEG compression is actually a complex, six-step processing procedure. Moreover, it is a parametric algorithm: one can adjust the loss rate depending on one's needs. But we have to be careful with this: it leads to a loss of quality. Other non-destructive compression algorithms exist, but the resulting images are more cumbersome and they may be unrecognized by some browsers.

Here are the three main traps to avoid with JPEG images.

- A JPEG image contains at most 256 levels per color (8 bits per color), that is 16000 colors. The compression leads to a posterization effect.
- Loss of accuracy: details are averaged in some way, resulting in blurred images with artifacts which can be misinterpreted.
- An image experiences some degradation every time it is saved as JPEG (a regular image file copy with a file manager does not affect the compression).

To avoid degradation once the final image has been processed, it must be saved in a non-destructive format,[28] and then a reasonably compressed JPEG copy can be shared on the Web. Saving in progressive JPEG mode is of interest if the image is to be accessed by a browser with limited bandwidth: the blurred image is quickly displayed at first, then the image progressively recovers its full accuracy as the browser reads the remainder of the file. Other types of compression exist, but they are more rarely used (JPEG 2000, PNG).

2.10.2 TIFF uncompressed images

TIFF (Tagged Image File Format) image format is widely used by professionals. It manages sixteen million (16 bits) to four billion (32 bits) colors. It allows a limited but non-destructive compression. The file size is substantially larger than JPEG – too large for the Web – and we have to remember that still cameras have a limited storage capacity. The format also accepts various encoding standards for professional use (CMYK, YcbCr, and CIE lab). The uncompressed 16-bit TIFF format is perfectly adapted to lunar imaging; in addition it is supported by almost all image-editing software.

2.10.3 FITS astronomical-format images

FITS (Flexible Image Transport System) was especially developed for astronomy, primarily for the Hubble Space Telescope. It is an uncompressed format, able to manage 16 bits per color (sixteen million colors), 32 bits per color (four billion colors), and a particular, floating-point, 32-bit format (which is variable but of extremely wide range). Moreover, it can embed text data to keep information about shooting parameters and any additional user data (comparable to EXIF data).

This evidently far exceeds our needs for lunar imaging, and it is exclusively managed by astronomical image-processing software (e.g. PixInsight, Fitswork, IRIS ...). But NASA then developed and provided a free Photoshop plugin called FITS Liberator. It is now autonomous and allows the conversion of an image from or to FITS format prior to processing with any image editor:

http://www.spacetelescope.org/projects/fits_liberator/

Moreover, a handy image-file manager such as XnView recognizes it (as long as the file type is enabled in the user preferences). The workflows in Section 7.2 show convenient ways to exploit the FITS format for lunar imaging. Once the processing has been completed, the final image can be converted into a more standard format, such as TIFF, in addition to a compressed format for the Web, such as

---

[28] Non-destructive compressions exist (e.g. in TIFF format), but the compression is not very efficient. Other types of popular, non-destructive compression software are listed in Appendix 2.

JPEG. It is highly recommended to keep the unprocessed, stacked, FITS image as a backup.

2.10.4 Raw formats (NEF, CR2, MTS . . .)

These formats are not intended for direct processing. NEF and CR2 are raw formats for Nikon and Canon DSLRs, respectively. They are not (Canon) or only barely (Nikon) affected by in-built pre-processing to diminish noise (offset and automated subtraction of hot pixels), and they preserve the genuine 12- or 14-bit conversion from the camera. The software provided (and third-party software libraries like DCRaw) can convert these proprietary formats to standard formats with no loss in quality, with Bayer matrix decoding, and, of course, the converted images contain all the noise and offset which are normally subtracted prior to internal conversion to JPEG format aboard the camera. This is very important for deep-sky image processing because the raw image remains unaffected. Lunar image processing is not so involved because the images mostly contain large, bright areas with a high signal/noise ratio, while deep-sky images contain mostly a restricted range of shades with a limited signal/noise ratio (with the exception of stars).

The MTS (or M2TS, or M2T) extension is for the AVCHD movie format exploited by high-definition camcorders and DSLRs in video mode. It is designed to preserve accurate images in a continuous stream, but it is poorly adapted to direct processing. The best approach is to convert the AVCHD movie into individual frames (using some of the freeware mentioned in Appendix 2). Nonetheless, frames from an AVCHD movie suffer from destructive compression, even at the highest flux.

2.10.5 Video formats for webcams

While video editing has countless formats, only one is useful for astronomical imaging with a webcam: the widely used AVI (Audio Video Interleave) format. It is only a container, and many compression algorithms may be concealed in an AVI file (for instance XDiv, DivX® . . .). Recording a movie is a good way to store numerous frames in order to process them in the aftermath of an observing session. In the recent past, a maximum of two gigabytes was the limit; but this has been surpassed by recent versions of popular freeware like FireCapture, RegiStax, and others. However, this represents a huge number of frames, while 200–1000 frames are enough for numerous lunar subjects. We have to be careful when we start to record an AVI movie: the frames must remain uncompressed. This results in larger files, but compressed frames are greatly altered and they are not at all suited to astronomical image processing. This selection is chosen in a check box in the webcam acquisition software menu. In addition, webcams may implicitly compress images prior to recording them, in order to maintain a high frame rate. It is wise to perform tests by recording the scale of a ruler or a calibration chart to determine the best FPS rate with no compression.

Of course, an AVI movie is supposed to be in color, while lunar imaging is often in black and white. If we choose to record monochrome images with a color webcam, the saturation is set to zero but this does not reduce the uncompressed file size. We can note that an AVI file is advantageously compressed for archiving and transfer by popular compression software/freeware (Appendix 2), with compression rates close to 50%. This does not alter the frames.

All other movie file formats are unsuited (or unrecognized) for lunar image processing because of the compression (even excellent ones like the H264). MPEG and MOV formats are widely available, but they have proved not to be appropriate for image processing.

2.10.6 DSLR video modes

Standard AVI video is one of the best choices, but it may include an implicit compression. MOV video has been used successfully, essentially because it can be converted into AVI for stacking, or directly stacked with AutoStakkert! 2. HD videos like 720p or 1080p are compressed into AVCHD or MPEG-4 (H264) formats (see Figure 2.25). The highest speeds are possible in an interlaced mode like 1080i, but interlaced modes are extremely prone to turbulence. No differences have been reported since 2010 between the final images from DSLRs and webcams or entry-level planetary cameras. As of 2016, few lunar photographers rely on a DSLR in video mode, even if an SD card now has enough room to store several tens of gigabytes. The main problem is that AVCHD movies do not contain raw images. Some DSLRs offer a cropped video mode with a limited but non-interpolated VGA resolution at 60 FPS (e.g. Canon EOS 550D, Nikon D60), which should be compared with the 397 FPS rate of a more recent ASI174MM/C planetary camera at the same resolution. Another solution is recording LiveView raw frames with a computer through a USB port.

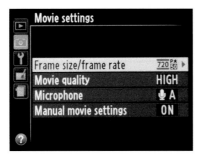

Figure 2.25 Setting the movie type of a DSLR: frame size (720 vertical lines), progressive (P) mode to avoid frame interlacing, FPS (50 frames per second), and image stream quality (high) to determine the loss in quality due to compression.

However, some DSLRs have an uncompressed HDMI output to record raw movies with a field recorder. The limit of 24 or 29.97 FPS is temporary, and DSLRs are likely to reach 100 FPS in the future. The best solution for now is to record uncompressed movies via the HDMI output to an external recorder like the classic BlackMagic HyperDeck Shuttle, the Atomos Ninja, or any other comparable portable and autonomous recorder to store the movie with no compression. The original, 12–14-bit frames are converted to 10 bits anyway. The remaining step is transcoding the movies (in ProRes or DNxHF format) into FITS or TIFF frames for future stacking and processing.

2.10.7 SER astronomical video format

Specifically developed for industrial/astronomical cameras, this movie file type contains uncompressed frames. It was initially devised for Lumenera and DMK cameras with Lucam Recorder software acquisition by German programmer Heiko Wilkens. The format is now recognized by major astronomical image-processing freeware and is not for viewing; it is a kind of FITS-format movie. Particularly interesting are embedded data (such as FITS files), including room to record shooting conditions (telescope, observer name ...), and time stamps. The acquisition software uses the system time (computer clock), and then saves time stamps into the SER movie. The embedded date/time data are no longer affected by file transfer, for instance from a computer to an external disk, and do not stick to the file parameters affected by operating systems' file managers. In other words, if we forget to include the date and time in the file name, the SER movie does it for us. This is particularly useful for quick, rapidly evolving phenomena such as occultations or transient lunar phenomena. The start and end date/time data in addition to the number of frames are sufficient to recalculate the time stamp for each frame with a good accuracy (8 bits for the date, 8 bits for the time). Some implementations theoretically record the time stamp with each frame. In addition, some image-acquisition software (e.g. FireCapture) can record a text file including the time stamp and other technical data for each frame. The SER format saves a lot of hard-disk room, especially if the camera is monochrome (60% smaller than a color AVI). Version 3 can record color images, but some software can manage only the monochrome version (as of 2016). This has few consequences for lunar imaging, because lunar color (or multispectral) imaging is advantageously feasible with monochrome cameras.

2.10.8 EXIF data

EXIF (Exchangeable Image File Format) is a data standard to embed shooting information in some image-file formats such as TIFF and JPEG (with the exception of JPEG 2000). Data are automatically written by compact and DSLR cameras: the camera model, the photographer's name (if it has been set with the help of the camera menu), date and time (local time and time zone), exposure, ISO, aperture, GPS coordinates (if the camera is GPS-equipped), and so on. This is extremely useful if we want to exploit HDR images, because dedicated software directly uses the information. Nonetheless, we have to be extremely careful when we use image-management/editing software because some types of software "forget" this information while others rewrite it! This implies performing trials before adopting an image-processing workflow with given versions of software.

The astronomical FITS format also contains a large header to include data in addition to the image itself. Some image-management utilities (e.g. XnView) allow the storage of additional data, but they are not recognized by other software. The advice is to write separate text files with the same name and a ".TXT" extension to keep the information safe, or to include the main data in the file name (Section 12.2).

2.11 Camcorders

Camcorders are valuable tools for recording family souvenirs, but they are not designed for astronomy (an example is given in Figure 2.26). The improvements of the sensors cannot offset their main drawbacks: the impossibility of swapping objective lenses,[29] weak zooms, and automated operation with very little and poorly ergonomic freedom of setting. Astronomical shooting requires manual

Figure 2.26 A cropped and strongly magnified image from an excellent consumer HD camcorder, with the optical zoom set to 10× and the optical stabilizer on. The limited manual setting of exposure and iris was unable to avoid overexposure. The red arc results from atmospheric refraction and chromatism.

---

[29] Except for semi-professional/professional camcorders, which will be mentioned later.

setting for exposure and focusing. Viewed through a camcorder, the Moon is generally reduced to a bright, tiny disk on a dark background: a situation that automatic setting is unable to correctly manage. Moreover, astronomical subjects demand extremely high-quality optics.

Fortunately, camcorders are able to shoot planetary and lunar conjunctions, or lunar eclipses, as long as a part of the image consists of a moderately light landscape: this is the only way to guarantee a correct, average exposure, with a correct automatic focusing. The problem is the same for miniature video cameras and compact still cameras.

Lunar eclipses should be a perfect subject for camcorders, because the continous variation in brightness should be automatically compensated for by the automatic gain control (AGC; see below). Unfortunately, a field test proved that the Moon remains overexposed. As with compact cameras in automatic mode, both automatic exposure and automatic focusing fail unless a large part of the framed scene includes a sufficiently well-lit landscape. Moreover, the subject requires a powerful zoom. Another solution is digiscopy with a telescope, bringing a comfortable magnification so that the Moon occupies a large part of the field.

Owing to the convergence of still cameras and camcorders, both are now able to record high-quality movies. Camcorders offer more storage, but high-quality still cameras set in video mode offer better optical quality, more powerful zooms, and manual setting functions. Under such conditions, no one should be surprised if, for the same cost, still cameras are increasingly preferred.

All camcorders have an automatic mode, a functionality called AGC. The device automatically sets the exposure/iris according to the ambient lighting. This is extremely useful for everyday movies, when technical concerns are not important. But, since this automation is not disengageable in most cases, we cannot manage the enormous brightness gradient of a scene comprising a shiny little spot lost somewhere in a very large and dark background, as occurs during a lunar eclipse. Even in "spot mode," the measurement and the automatic setting of exposure by the AGC results in an overwhelmingly overexposed lunar disk. A few camcorders offer a tiny amount of freedom for manual exposure setting. Furthermore, we must acknowledge that camcorders' ergonomics is often disastrous: a structured menu, controlled using microscopic buttons placed on a mobile, plastic panel attached to the body by a fragile joint, is not handy, especially at night.

Figure 2.27 illustrates the crippling effect of the AGC. We could imagine that a neutral-density filter, attached in front of the lens, would dim the Moon enough for one to see maria and craters. But the AGC reacts and increases the gain, thus the image is affected by noise and overexposure. The only way to fully control the indispensable settings (especially exposure and iris) is by using a semi-professional camcorder, e.g., at present, Sony HandyCam/HXR series, Panasonic AG-AC series, Canon XF series, JVC GY series, and many others. In addition, some have interchangeable lenses, which are compatible with DSLRs.

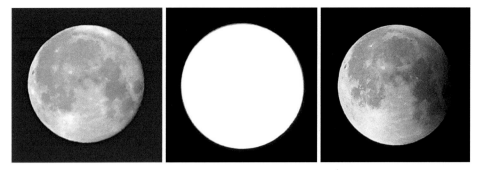

Figure 2.27 Single cropped frames from another HD camcorder with optical zoom set to 20×. Left: in daylight, very close to the horizon (15°), the Moon was favorably dimmed by the atmospheric absorption, while the background sky was not black, resulting in a well-balanced automatic exposure. The brightness was set to –2. The color balance was not properly set. Center: when the Moon was higher, the image was strongly overexposed, despite the "brightness" being manually set to the minimum value of –6. The freedom in setting parameters was too narrow to mitigate the enormous local dynamic of the scene. Right: a frame captured during the penumbral step of an eclipse when it was 16° high, the AGC was active in "spot" mode, and the Moon was very precisely centered with the help of a motorized mount. Any de-centering resulted in strong overexposure.

2.12 Webcams

A webcam is the most affordable technology and a very efficient means to shoot the Moon in detail. Owing to its small sensor, a webcam has a very narrow field, but this is not related to the final resolution. A webcam can be exploited with any telescope in prime-focus imaging, eyepiece projection, or Barlow projection (see Figures 2.28 and 2.29). We saw in Figure 1.7 that a webcam can also be simply attached to a basic beginner's telescope with adhesive tape for prime-focus imaging. The next step is to find the Moon with the help of the viewfinder. The latter has two, three, or six screws to align it in parallel with the telescope. This alignment is easily done in daylight, with the help of a TV antenna or a chimney.

As most webcams have cheap, CMOS sensors with a slow FPS rate, successive images at high magnification may be distorted by the combination of possible vibrations, turbulence, and the rolling shutter. The telescope must be stable to attenuate the motion-blurring effect from vibrations. A solution is to reduce the "resolution," that is, the effective number of pixels, for example from 1.4 megapixels down to a reasonable 300 000 pixels (VGA resolution). This increases the number of frames per second, and hence there is less distortion caused by the motion blurring from one frame to the next. This can be neglected if the instrument is stable and the magnification is moderate.

The image-acquisition software is provided by the supplier, but FireCapture and other software can easily manage webcams with the help of the standard

Figure 2.28 A webcam simply placed at the prime focus of a Celestron 5 had photosites that were too large to record all the details of the cast image, resulting in thickened crater rims with pixelization (30° E, 31° N).

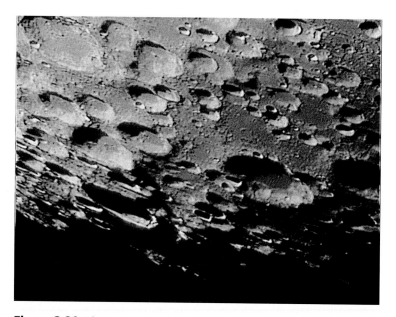

Figure 2.29 The same webcam with a stronger magnification due to eyepiece projection. The telescope was larger but the most important thing is that the tiniest details of the optical image cast on the sensor were commensurate with the photosites of the camera (19° E, 67° S).

WDM (Windows Driver Model) software interface. Whatever the acquisition software, we now have to determine the correct parameters. This is quite simple.

- The exposure determines both the electronic shutter-release duration and the FPS rate. For instance, choosing 100-ms exposures implies 10 FPS. We have to empirically limit the FPS because of limitations of the USB interface and an implicit compression by the camera to overcome the bandwidth limitation at the price of a loss in image quality.

- The color balance must be set to automatic, then we wait for it to stabilize for some seconds, and finally it must be unchecked for repeatability during acquisitions. Choosing the black-and-white mode is useless because the driver does not deliver undebayered images.
- The movie format has to be uncompressed AVI for further processing.

Webcams generally offer up to 1.2 megapixels, enough to show a final image having the same dimensions as the entire screen of a laptop. The main drawbacks of webcams are as follows.

- Lower resolution and less sensitivity due to the color sensor (as with most consumer imaging devices).
- Low frame rate at high resolution because of limited throughput. In practice, excellent lunar images were taken with 640×480 pixels at a modest frame rate of ten images per second.
- Interpolation, that is, computing artificial intermediate pixels to simulate a higher resolution. Actual and interpolated resolutions are mentioned on the packaging. Astronomical imaging requires non-interpolated images; this has to be set in the webcam's application software (select "resolution" and choose the maximum actual resolution).
- Possible problems in adapting them to telescopes. Reliable solutions are commercial adapters (12×0.75 thread lens-side, 1.25-in or T2 eyepiece holder-side). They are widely available from many dealers nowadays. But many webcams are not intended to operate without the built-in lens, and perhaps we will have to carefully open the plastic housing with a small screwdriver. **This voids the guarantee**. The lens must be loosened counterclockwise and then removed, to be replaced by the adapter. Opaque adhesive tape can cover the LED, if any, to eliminate a parasitic lighting source.

It is best to exploit the original acquisition software even if the operating system prompts the use of its own utilities. Original software usually lets us determine essential parameters: manual exposure, color balance, FPS, and so on. Still images can be directly captured by the software, or extracted from the AVI movie (e.g. VirtualDub, ImageJ). AVI movies can be processed with the help of numerous types of astronomical software and freeware, e.g. AutoStakkert, AviStack, IRIS, and RegiStax.

Discontinued, but still exploited and present on the used-equipment market, the Philips CCD webcam family (VestaPro™, ToUCam™, ToUCam Pro™ ...) can be flashed up into raw modes at 10 FPS with no compression. The modification is reversible. As of 2016, the drivers have not been maintained for more than a decade. Some Philips CCD webcams have been flashed to run with more recent operating systems, others will need some effort. A good example is using "virtual machine" software to run, when needed, an older operating system with the original software of the webcam. We can find this (VirtualBox, VirtualPC, QEMU, WINE, VMWarePlayer ...) on major home operating systems (Unix, Windows, OsX ...), or use a bootable USB key with Windows XP.

Some "planetary" cameras are nothing more than modified webcams, with a solid housing, no lens, adapted software, no artifacts due to onboard pre-processing, and no image compression, and they are provided with an adapter for telescopes. This is why we ought to consider the direct purchase of an entry-level planetary camera rather than a webcam.

2.13 Analog-output video cameras and electronic eyepieces

These are primarily video-surveillance and industrial cameras (Figure 2.30). Consumer miniature cameras are fully automatic, with a non-disengageable AGC. The lens is removable by unscrewing it, as with most webcams. Electronic eyepieces are miniature cameras with a 31.75-mm (1.25-in) adapter to be directly inserted into the drawtube. All have an analog video output, either composite (image and synchronization are mixed) or S-Video for a better quality. Professional analog-output cameras may have separate signal components (red, green, blue; horizontal and vertical synchronization; sometimes a synchronization signal is mixed with a color). For historical reasons, the resolution is often limited to 576 horizontal lines (CCIR cameras).

There are three types of optics threads:

Figure 2.30 Left: the rear side of an analog-output, professional video camera. In the bottom corner of the housing on the left there is an S-video (labeled here S-VHS) socket for better image quality. Opposite this there is a composite output (all video signal components are mixed) with a BNC connector; here a more widespread RCA/Cinch connector adapter is plugged in. Right: the thread for C or C/S optics is compatible with 31.75-mm (1.25-in) or T2 adapters.

Figure 2.31 A miniature video-surveillance camera. The removable lens can be replaced by a 12×0.5 adapter (like a webcam) to T2 or 31.75-mm (1.25-in) for prime-focus, Barlow, or eyepiece projection imaging. Connectors from left to right: power supply, microphone (not very useful for astronomy), and composite video.

- 12×0.5, like webcams (the same $12 \times 0.5/31.75$-mm (1.25-in) or $12 \times 0.5/\text{T2}$ adapter may be installed in place of the lens; see Figure 2.31)
- C mount
- C/S mount

The C and C/S mounts differ in terms of their lengths (C/S stands for short C, denoting recent, high-quality miniature video-surveillance cameras). A 5-mm spacer may be inserted to convert a C/S mount into a C mount. Numerous planetary cameras are fitted to C or C/S threads, and adapters are widely available. Old 16-mm film cameras' lenses are directly compatible with C mounts; unfortunately, their optical quality is inferior to the necessary astronomical quality, as is the case also for most ordinary video-surveillance lenses. Professional C-mount lenses are of excellent quality, but they are quite expensive (e.g. TechSpec®, professional-range Schneider, Zeiss, and others). This is why the cameras are generally fitted to a telescope with a C/T2 or C/31.75-mm (1.25-in) adapter.

Frames are to be recorded by an external device, either an analog or a digital recorder (Figure 2.32), or a computer. Some camcorders have an S-video or composite input, but this is rare. The image can be directly viewed on a TV screen or a video projector by using the S-video or the composite input. The best way is to convert the analog stream into digital frames via a video grabber (Figure 2.33).

Since the exposure setting is fully automatic, the camera constantly attempts to establish an impossible balance between the darkest and the brightest areas of the scene (Figure 2.34). If the main part of the image is dark, the Moon is overexposed, and vice versa. The worst situation is a thin crescent: while more than 90% of the

Figure 2.32 An old-school, but handy and autonomous, Sony Video Walkman to directly record movies from analog-output cameras in the field.

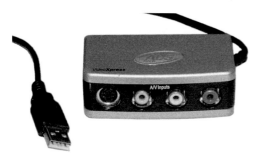

Figure 2.33 A USB analog-to-digital video converter, informally named a "grabber." From left to right: S-video, composite video, and stereo audio connectors. This is a better solution than photographing the screen! This is an 8-bit conversion, consumer model.

image is black, the crescent is totally overexposed. The only solution is to frame the Moon with the illuminated landscape to get the ideal balance between bright and dark parts in order to force the AGC to set an acceptable exposure.

This issue does not exist with professional cameras offering a manual setting. Successful views were obtained at moderate magnifications with an interlaced Mintron 12V1CV-ex designed for low-light video surveillance. Manual setting is done with an optional remote control or with the buttons on the rear side. The menu is on-screen display (OSD). Similar results are to be expected from comparable professional analog-output cameras from Vixen, Gstar, Orion, and others.

Figure 2.34 Saturn passing behind the Moon: image taken in 2001 with the miniature video-surveillance camera at the prime focus of a telescope with a resulting focal length of 790 mm. The right edge of the rings is truncated by the limb of the Moon. The brightest areas of the Moon were unavoidably overexposed by the AGC. The camera was simply held by hand behind the telescope. The image was recorded with a Sony Video Walkman in Video 8 format, then digitized with a grabber.

Here are some common characteristics of these systems.

- Number of pixels: depending on both the camera and the grabber (analog-to-digital image converter), often limited to 480 or 576 lines (vertical).
- Generally interlaced sensor (for compliance with the former broadcast video standard).
- Current supply: the interface does not supply the camera; it needs an external power supply, often 12 V DC. The current drag depends on the camera (<1 A).
- Connectors:
  - one RCA/Cinch connector for analog, composite video signal output;
  - one S-video connector for higher-quality, analog, Y/C video signal output;
  - three or more RCA/Cinch connectors for analog, component video signal output with separate links for each color and one or two links for vertical and/or horizontal synchronization.
- The video signal may be PAL, SECAM, or NTSC. PAL signal is to be favored (this was the former standard for analog broadcast equipment).

2.14 Digital still cameras

2.14.1 Compact and bridge cameras

These cameras differ by their size, their freedom regarding manual setting, and their sensors. Compact cameras have a very small sensor with numerous pixels, so the small photosites (e.g. 2.5 μm or less) and wells provide limited performance.[30] Nonetheless, high-end cameras in each category tend to be fitted with the larger-format sensors of the superior category.

- Entry-level compact camera: sensor diagonal is 7 mm.
- Bridge camera: sensor diagonal is 10–21 mm.

A 50× optical zoom is sufficient to shoot craters. A digital zoom (a software interpolation) is useless, and is tending to be abandoned by manufacturers. If the exposure lasts a large fraction of a second, motion blurring is to be expected. The motion blurring hardly comes from the diurnal motion itself because the Moon is bright enough to guarantee short exposures: it rather comes from the photographer, and this necessitates a tripod. In contrast to most astrophotographical subjects, an ordinary, slow F/D ratio (e.g. F/D = 8) at maximum zoom extension is sufficient.[31]

The first thing to do is to neutralize the flash, unless the flash is powerful enough to illuminate the Moon at a distance of 380 000 km! A special, quick-access button is often placed at the rear of the body to stop it. The camera compensates for the low mean brightness of the scene by increasing the sensitivity (that is, a software gain, which is harmful for the signal/noise ratio), but, more interestingly, the exposure duration is extended. Some cameras offer a specific "magnificent night scene" mode with no flash and long exposures. This is useful when the Moon shows a thin crescent, earthshine, or a conjunction.

The main traps are automatic exposure and automatic focusing.

- Automatic focusing is based on the detection of close, large objects in the field of view with either infrared or ultrasound reflection, phase contrast, or contrast measurement. Since the Moon is located at optical infinity, the camera acts as if it were a distant landscape. This proves to be poorly reliable because the global contrast is unevenly distributed: a small, bright spot on a large, black background is not a normal scene for a consumer camera. DSLR, bridge, and compact cameras often forbid shooting when they are unable to focus, indicating that the subject is too dark. A first possible solution is to manually focus after having set the AF/MF button (close to the objective) to MF (manual focusing, rather than automatic). Many low-end cameras do not allow this option. A second solution is to re-center the scene, so that a large portion of sufficiently lit, distant landscape allows the camera to focus to optical infinity.
- Automatic exposure and automatic iris have the same problem because of the uneven distribution of light in the scene. The Moon is often overexposed,

---

[30] In an extreme case, the 16-megapixel sensor of the Galaxy S5 smartphone has minuscule, 1.12-μm photosites, but the BSI technology compensates for losses at the expense of dynamic range. This is to be compared with a deep-sky camera with 6.5–20-μm photosites.

[31] The F/D ratio is always indicated at the front of the objective, e.g. 1:2.8–5.6 means that the minimal focal length is 2.8 times the aperture, and, at maximal extension, the focal length is 5.6 times the aperture: less brightness but more magnification.

leading to excessive contrast, except in daylight. By night, manual exposure setting is to be preferred. In automatic exposure mode, re-centering the scene to frame a part of the landscape is, once again, a solution that can be employed to achieve a correct, global exposure.

Some other details condition the adaptation of the camera to lunar imaging.

- The size of the screen is important for comfort and to ensure that focusing is correct.
- An orientatable screen is a real advantage when we are aiming at the Moon overhead.
- Bridge cameras often have a thread around the objective for digiscopy.
- Displaying a clear histogram is very important to avoid overexposure, because a simple control on the uncalibrated screen is not reliable.
- Ergonomic aspects should not to be neglected: the camera has to be used in very-low-light conditions. A flashlight is very handy to locate the menu buttons.

2.14.2 Hybrid and DSLR cameras

Apart from the sensor size, the fundamental difference between these types of camera and compact/bridge cameras is the determining advantage of removable optics. This means that, in addition to the compatibility with various telelenses, the body can be directly attached to any telescope (see Figures 2.35, 2.36, 2.37, and 2.38).

The main difference between a hybrid and a DSLR (digital single-lens reflex) has long been that the viewfinder of a hybrid camera does not really show the image cast on the sensor. Today, the difference has disappeared: both DSLR and hybrid cameras no longer have optical viewfinders, and the mirror of a DSLR is an endangered species: the image cast in the sensor is displayed directly.

Figure 2.35 This image is severely cropped (it amounts to less than 4% of the original image!). It shows a tiny Moon with maria and some large craters toward the South pole. This 105-mm zoom is not powerful enough to reveal smaller details. Nevertheless, after a software magnification by a factor of twenty-five(!), the image is still acceptable despite the entry-level, 24-megapixel APS-C CMOS sensor.

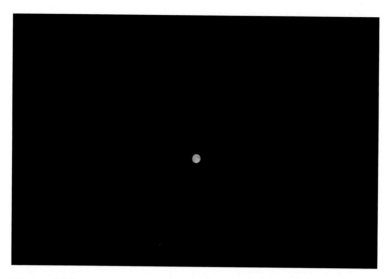

Figure 2.36 Uncropped image with a 55-mm focal length.

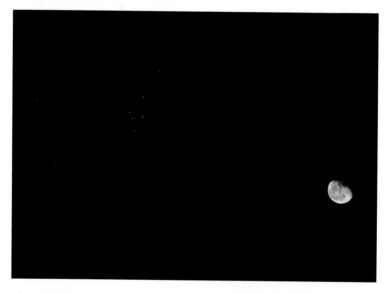

Figure 2.37 Uncropped image with a 105-mm focal length.

The sensor size is always more than 10 mm in the diagonal, and this greatly enhances the signal/noise ratio and the dynamics:

- entry-level DSLRs and quality hybrids: 28 mm (APS-C)
- professional DSLR: 43 mm (full-frame)

This is to be compared with professional, medium-format cameras, which can have 80-mm-diagonal sensors.

Figure 2.38 The same DSLR as for the previous three figures is here directly attached to a telescope of focal length 1250 mm in place of the photolens.

A 300-mm telelens is powerful enough to shoot large craters. Owing to the difference in size relative to compact cameras' sensors, this focal length leads to a lower magnification, but the quality is far better. The main limit is the diurnal motion.[32] The Moon may remain in the field for only some tens of seconds, depending on the focal length and the size of the sensor. An equatorial mount is a serious advantage, not only because it can compensate for the lunar motion, especially during a lunar eclipse, but also because precise and smooth alignment greatly facilitates framing and tracking at high magnification. Figures 2.35 to 2.38 depict how different optics attached to the same DSLR result in extremely variable magnification and field of view.

2.14.3 Dimension of the Moon with compact and DSLR cameras

The size of the Moon in the image depends on the focal length and the size of the sensor. Since the full Moon spans about half a degree (it is masked by our thumb, arm outstretched), it has to be substantially magnified. Tables 2.10 and 2.11 show the dimension of the Moon on the image with respect to different sensor sizes and focal lengths (see Figure 2.39) for a DSLR with an APS-C sensor (approximately 22 × 15 mm) and for a compact camera with a 1/2.3-in sensor (approximately 4.6 × 6.2 mm).

To interpret these tables, we have to know two values: the size of the sensor, specified by the manufacturer, and the actual focal length while shooting (because

---

[32] In addition to the proper motion of the Moon (Section 1.2.1).

Table 2.10 Field of view and relative size of the Moon on an APS-C sensor with different photolenses and telelenses. The last lines of the table correspond to an astronomical instrument.

Focal length (mm)	Field diagonal (°)	Moon size as percentage of the vertical field (%)
5	138.9	0.4
10	106.2	0.7
18	73.0	1.1
25	56.1	1.5
35	41.7	2
50	29.8	2
85	17.8	5
105	14.5	6
180	8.5	10
300	5.1	17
400	3.8	23
500	3.1	29
800	1.9	46
1200	1.3	70

Table 2.11 Field of view and relative size of the Moon on a 1/2.3-in sensor with different photolenses and telelenses

Focal length (mm)	Field diagonal (°)	Moon size as percentage of the vertical field (%)
5	58.4	1.5
10	31.2	3
18	17.6	5
25	12.7	7
35	9.1	10
50	6.4	14
85	3.8	23
105	3.0	29
180	1.8	50
300	1.1	83

a zoom has a variable focal length), which is displayed on the screen and stored in EXIF data.

2.14.4 High dynamic range (HDR) by bracketing

Some sensors offer a genuine HDR/WDR capability, but a camera may simulate it by taking several frames at different exposures (two to seven, depending on the camera) within a very short time interval. The frames are combined, and the

Figure 2.39 A hybrid camera with a 24× zoom. This factor of 24 is the difference between the shortest and longest focal lengths. The actual focal length is indicated at the top of the lens (4.5–108) along with the F/D ratio (1:2.8). The latter may be variable. The 24 × 36 equivalent focal length is mentioned on the side of the body (25–600 mm).

resulting image may represent a 20-bit image in terms of dynamic. The camera may take 0.5–12 frames per second, but some use a video mode and reach 60 frames per second with a limited number of pixels. The range in exposure values varies from 2 (a very limited HDR) to 18 for extreme values.

If the camera is adapted to a telescope, in contrast to the case of quick HDR and WDR sensors, HDR cameras basically use a classic, multi-bracketing function, requiring a certain interval between one frame acquisition and the next, which may result in image distortion by the turbulence. In addition, some DSLRs release the mirror for each frame, resulting in vibrations (see Section 2.14.8). Despite the drawbacks, the function is useful at low and moderate magnifications, because the turbulence is limited and the contrast is maximal, especially when the Moon is gibbous or full. At high magnifications, the contrast decreases substantially, and an HDR capability is less important, if at all.

In case our camera does not offer the HDR functionality, the procedures for multiple-exposure combination are mentioned in Section 8.11.

2.14.5 Choosing the sensitivity and exposure

According to a long-standing tradition among film photographers, the sensitivity of a digital camera is rated in ISO, after the film-sensitivity scale. Commonly available films were rated from 25 ISO (e.g. Kodak Ektar 25) to 3200 ISO (e.g. Scotch Chrome 800–3200). In a similar way, a digital sensor has by construction a fixed sensitivity. By having twice the sensitivity, we can set half the exposure, giving less turbulence blurring. But setting the sensitivity on a compact, hybrid, or

Figure 2.40 Setting the exposure and the aperture. The sensitivity is here set to 200 ISO because the subject is bright (the gain is simply lowered internally). Here "1/80" is the exposure duration in seconds, while "F5.6" is the F/D ratio. This menu seems complex, but an opportune "?" icon means that a help button on the rear of the camera body explains the parameters. According to the brightness measurement (spot, weighted . . . depending on the camera), the exposure is correct when the "o" is centered on the horizontal scale. The iris is here set to the –1.3 notch to mitigate the risk of clipping.

DSLR camera means lowering (e.g. to 100 ISO) or increasing (e.g. to 6400 ISO) the lightness of the image *after* it has been recorded by the sensor. This process – the gain – is automatically performed in the camera. Common, APS-C DSLRs have a real sensitivity of about 800 ISO. The recommended process is to adjust the necessary gain with specialized, image-editing software rather than letting the camera do it in an internal, unverifiable way. The best sensitivity varies according to the sensor. To determine it, we can take exactly the same picture (at finest resolution) with 200 ISO, 400 ISO, 800 ISO . . ., dividing the exposure duration by two, e.g. 1/50 s, 1/100 s, 1/200 s (Figure 2.40) . . ., and then choose the best image in the sorted series, that is, the one with as much brightness as possible but as little noise and blurring of detail as possible.

Here is an example of determining the best sensitivity, along with other artifacts from the camera's internal processor, by Christian Buil:

http://www.astrosurf.com/buil/cameras.htm

Once the image has been taken, it is essential to verify the exposure (Figure 2.41). The worst case is overexposure because the image cannot be fixed afterwards. Since the camera screen is small, often badly orientated (but some are orientatable), and uncalibrated, the only safe way is to display the histogram. If the curves do not reach the right part of the diagram, no overexposure happened. The leftmost part of the curve mixes the lowest levels of the Moon and the sky background.

2.14.6 Setting the image type

Amateur photographers favor JPEG images because they save a substantial amount of storage capacity and are directly readable by all image file managers. Raw images

Figure 2.41 The histogram displayed by the camera. Top: the luminance. Below: the three fundamental components, red, green, and blue. Of course, the Moon is globally (but not totally) monochrome, and the three curves resemble each other. If the curves reach the right part of the grids, the image is definitively overexposed. Some cameras display a luminance curve only, or superimpose the three colored curves.

Figure 2.42 Choosing the image type in the camera menu in two different DSLRs. The image can be simultaneously stored in JPEG and raw formats. Raw images have to be converted by the proprietary software provided, or by third-party software with specific libraries, but they preserve the 12- or 14-bit accuracy from the sensor, with no artifacts from compression and noise reduction by the processor aboard the camera. Fine/normal/basic and L/M/S (large/medium/small) correspond to the number of pixels.

preserve – to a greater or lesser extent, depending on the manufacturer and model – the original frames. Figure 2.42 shows two examples of image type selection.

Another parameter is the number of pixels. A 12-megapixel image may prove to be effectively equivalent to a 24-megapixel image on the same camera with an entry-level CMOS sensor. Experiments are useful to determine the best balance between the number of pixels (called "image quality" or "resolution" in the camera menu) and the size of the image file. The manufacturer may use a hardware binning, multiplying the sensitivity by grouping the photosites, resulting in a substantially better image despite the number of pixels being greatly reduced.

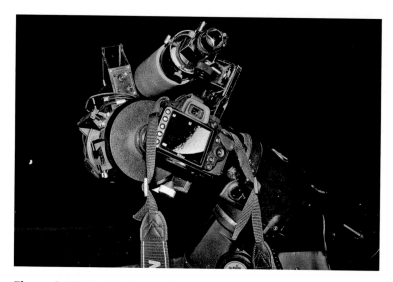

Figure 2.43 Live focusing on the screen of a DSLR attached to a telescope at prime focus. The Moon is one of the rare astronomical subjects which are bright enough for LiveView. Zooming considerably facilitates focusing. An orientatable screen is highly recommended unless the tripod is very high when the Moon is high, because the presence of a diagonal could excessively extend the distance to the focal plane. The lateral focuser of a Newtonian telescope fits perfectly to non-orientatable screens.

2.14.7   Focusing with the viewfinder or with LiveView

Focusing through the viewfinder has always been the bane of astrophotography. Film SLR viewfinders had frosted glass to cast the image, to ensure that the eye does not partially focus in place of the optics, at the price of a strongly darkened image. Modern DSLRs have transparent glass, which saves brightness at the price of the risk of incorrect, manual focusing, assuming that most photographers rely on autofocus. Since sensors can continuously display the image, modern DSLRs may display the live image from the sensor (LiveView, see Figure 2.43). Even if the camera lacks a continuous display, this can be simulated – albeit at a considerably lower speed – with the help of a computer and the DeepSkyStack Live freeware by Luc Coiffier (Appendix 2). It continuously monitors a folder where images are transfered from the camera by third-party software, to help focusing at a low frame rate. Cameras have an HDMI output which can be exploited for live streaming. The solutions are not mature as of 2016, but the situation is evolving quickly.[33]

[33] Experiments have been performed with Magic Lantern, Canon2Syphon, LiveStream Broadcaster, Ustream Producer, Black Magic Intensity Extreme … Owing to the stringent demands and the convergence of video cameras and DSLRs, live-streaming solutions are not stable yet, and the options are evolving rapidly.

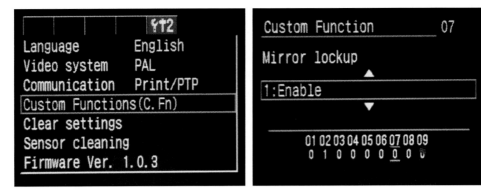

Figure 2.44 This is a part of the menu of a Canon EOS 350D. The "Custom Function" #7 allows the raising and locking of the mirror in advance. Then shooting is not affected by shock or vibrations arising from the mirror's motion. Raising the mirror and using a remote control are essential for lunar imaging with a DSLR. A mirrorless DSLR also requires a remote control for ease of use and comfort.

Whatever the solution is, be it LiveView, a classic viewfinder, or continuous monitoring of frame transfer, we have to find the best lunar location for focusing: a contrasted crater near the terminator, or, in the case of a full Moon, the limb. On gaining experience, our eye becomes able to perform a kind of real-time averaging of details through the turbulence.

2.14.8 Flipping the mirror in advance

Some DSLRs allow the raising of the mirror prior to shooting (and, of course, after focusing), as illustrated in Figure 2.44. The reflex viewfinder is neutralized. Raising the mirror in advance eliminates any vibration, which could easily be amplified by the magnification. This feature is very handy when the DSLR is attached to a telescope or a powerful telelens. If the DSLR lacks this option, shooting with a remote control, preferably a wireless one because the connectors are fragile, considerably limits handling vibrations. An alternate solution, albeit efficient only for long exposures, is manual closing with black cardboard (Section 5.5.5). Note that, even in LiveView mode, many DSLRs uselessly lower and then raise the mirror at each shot, causing vibrations.

2.15 Planetary and industrial cameras for the Moon

2.15.1 Planetary cameras

A planetary camera is an astronomical or industrial camera with some precise characteristics adapted to planetary or lunar imaging.

- Uncompressed frames with no interpolation.

- A sensitive sensor to limit turbulence blurring with the help of short exposures (sensitivity or quantum efficiency mentioned in the datasheet).
- A substantial throughput (some tens to some hundreds of frames per second).
- A small sensor to get the best, central zone of the image field.
- A removable lens with standard C, C/S, T2, or other thread for photographic optics or telescope adaptation.
- A great dynamic capacity (high saturation signal mentioned in the datasheet).
- A digital interface: USB or FireWire (use of the latter for this purpose is becoming rare), Gigabit Ethernet (rarely used by amateur astronomers because it requires more wiring and an additional power supply), or Camera Link interface (used by professionals only).
- The sensor is often primarily designed for video surveillance rather than for industrial applications: it is suited to low-illumination situations, with a lower cost and a rolling shutter.

The shutter type and the maximal exposure are not significant for our purposes. All planetary cameras are utilizable, but for lunar imaging there are two specific criteria.

- A lunar camera should offer a relatively large number of photosites because the Moon has a large surface.
- The monochrome version of a given camera is always more efficient.

Here are some typical examples[34] for beginners as of 2016. In order to be very affordable, they are equipped with a consumer CMOS color sensor:

- Orion StarShoot USB Eyepiece Camera II, 1/4-in color sensor, 0.3 megapixels, up to 30 FPS, 6-µm photosites;
- Celestron NexImage, 1/4-in color sensor, 0.9 megapixels, up to 30 FPS, 3-µm photosites;
- Pentaflex/Perl Echorius/iOptron iE1300: 1/3-in color sensor, 1.3 megapixels, 16 FPS at maximal resolution, 30 FPS in VGA, 3.6-µm photosites;
- Orion Starshoot Solar System Color Imager IV: 1/3-in color sensor, 1.3 megapixels, 30 FPS at maximal resolution, 3.6-µm photosites.

The prices range from $70 to $150.

A better lunar camera preferentially has a USB3 interface, for a greater FPS rate, and a monochrome sensor, for more sensitivity and better accuracy (see the two examples in Figures 2.45 and 2.46). Nonetheless, recent (at the time of writing) color sensors are as sensitive as monochrome sensors of the previous generation, with a better sensitivity in the deep red and near infrared, and less noise. For instance, the Sony IMX224 and IMX185 color sensors can compete with the Aptina MT9M034 and SONY IMX104 monochrome sensors. As an atypical choice, a color

[34] Some cameras (especially Celestron) provide 10-bit or 12-bit frames, thereby reducing the maximal FPS, but with better conversion of shades.

Figure 2.45 This planetary camera has a T2 female thread, a removable C adapter, and a removable 31.75-mm (1.25-in) adapter, bringing compatibility with a very wide range of photolenses and all telescopes. The sensor is protected by a removable window, either with an anti-reflective coating (most monochrome cameras and some intentionally near-infrared-sensitive color cameras) or with an infrared- and ultraviolet-rejection coating (numerous color cameras). The black socket is the ST4 port used for autoguiding, which is primarily intended for deep-sky imaging, but the camera may be exploited as an autoguider for lunar time-lapse photography or lunar eclipses. The blue socket is for USB3. The rear side has a Kodak thread for a photo tripod.

Figure 2.46 This planetary camera has a C thread (shown with the 31.75-mm/1.25-in adapter), a USB2 port and an ST4 port (rear side). It also has a Kodak thread for a camera tripod in case the camera is fitted with a telelens. **Image by Etienne Martin.**

camera such as the ASI224MC performs well in "false monochrome," including in the near infrared with non-debayered images (Section 2.9.7). A color sensor is directly adapted to lunar geology, whereas a monochrome sensor needs a filter wheel.

Here are some examples for more advanced imaging as of 2016, with monochrome sensors.

87

- USB2/USB3 with limited FPS, previous-generation sensors, but still-efficient cameras: DMK 31/41/51, Celestron Skyris 445/274, I-Nova Plb-Mx/Mx2, QHYCCD QHY5 L-IIm, ZWO ASI120MM, Altair Astro GP Cam, and others.
- USB3 with high FPS, current- or previous-generation sensors, but improved noise reduction: QHYCCD QHY5III-174 M, QHY5III-290M, ZWO ASI174MM, AST290MM, Celestron Skyris 236M, and others.

Prices range from \$280 to \$680.

2.15.2 Industrial cameras

These cameras often exploit the most recent sensors intended for industrial applications (long exposures are not the main consideration; high speed and global shutters are favored). Some industrial cameras with high FPS rates and USB or Gigabit Ethernet interfaces are used by amateurs (e.g. Imaging Source's DMK, AVT Manta, Foculus, Basler ACA, Point Grey BackFly/Grasshopper3/ FLEA3, IDS μEYE, Lumenera CMV2000/CMV4000/LT425 NIR ...) with excellent results. These were the very first applications of recent sensors (during the years 2013–2015: IMX174, IMX236, ICX687, EV76C661, CMOSIS CMV2000 ...) in amateur astronomy. Some third-party software is compatible (e.g. FireCapture and Video Sky), and some suppliers have been developing their own software (e.g. Genika Astro at http://airylabs.com). Prices range from \$720 to \$2600.

The cameras are provided with standard mounts: C, C/S, T2, M48, or a DSLR-compatible mount (such as Nikon or Canon). The possible interfaces are FireWire (IEE1394.a/b), Camera Link (see Figure 2.47), USB2 or USB3, and Gigabit Ethernet. Some require an external power supply via the twelve-pin, Hirose connector (which is also used for triggering). Since a Camera Link grabber

Figure 2.47 The rear side of an industrial camera with a Camera Link interface (labelled DATA OUT) and the twelve-pin Hirose port (labelled DC IN/SYNC).

Table 2.12 Interfaces for digital-output cameras

USB (Universal Serial Bus)	The most common interface for general consumer devices. Transfer rate:
	USB1: 1.5–12 megabits per second (Mb/s)
	USB2: 480 Mb/s
	USB3.0: 5 Gb/s
	USB3.1: 10 Gb/s
	Current supply:
	0.5 A, sometimes less for low-end laptops. Up to 5 A with USB3.1 in recent desktops.
	Connectors:
	USB connector, A-type (flat) computer-side, A-type or B-type (rounded square) camera-side.
	A USB interface needs only a USB cable (with no hub or extension cable to prevent image degradation or software freezing).
GigE (Gigabit Ethernet)	Often used for industrial cameras for high transfer rate and networked cameras. Requires specific software drivers.
	Transfer rate:
	Up to 1000 Mb/s
	Current supply:
	Some cameras use POE (Power Over Ethernet) and do not need any external power supply, but most do.
	Connectors:
	RJ45 (square connector, inherited from computer networks, the same as home local area networks).
	Hirose (round connector with 12 pins, resembling an S-Video connector or a small DIN connector) for camera control and power supply (with the exception of some old Hirose connectors).
FireWire (IEEE 1394)	Designed for various, chainable devices (hard disks, digital video camcorders, cameras ...).
	Transfer rate:
	400 to 3200 Mb/s
	Current supply:
	Up to 1.5 A
	The four-conductor variant requires an external power supply
	Connectors:
	Several types (FireWire 800, 1394c variant, four- and six-conductor FireWire 400). Sony variant has four conductors and no power supply (it is integrated in the camcorder). Other common types of home equipment use FireWire 400 with six conductors (resembling a narrow, rounded square) for DV-camcorders. Cable length is 4.5 m at most.

Table 2.12 (cont.)

Camera Link	Designed for industrial and scientific cameras. Transfer rate: 2, 4, 5.4 or 6.8 Gb/s depending on configuration (base version with 24 bits, medium with 48 bits, full version with 64 bits, extended full version with 72/80 bits). Current supply: This interface basically does not supply the camera, except for the Power Over Camera Link (POCL) version. Connectors: The prescription is a 26-pin, MDR-26 (standard) or an SDR-26 (miniature) connector, but there are some proprietary connectors for the frame grabber (computer side).

meets professional requirements, it is available at a professional price, no less than $1200 with acquisition software.

2.15.3 Camera-to-computer interfaces

Table 2.12 presents a summary of available interfaces for digital-output cameras. Transfer rates are indicated in bits because cameras may produce various coded formats, from "bytes" (8 bits can represent 256 different levels per color or 256 levels of luminance for monochrome cameras) to "words" (16 bits for 65 536 levels), or any intermediate number of bits. Images are coded in 8–16 bits depending on the camera. Owing to technical constraints, any format greater than 8 bits is often transferred as a 16-bit format, halving the transfer speed. Mb/s means megabits per second; MB/s means megabytes per second. We express transfer rates in Mb/s in Table 2.12 because cameras may use any format (often 8–12 bits for planetary cameras).

3

Adapting your imaging device to the instrument

3.1 Digiscopy

3.1.1 Plan B: digiscopy (or afocal projection)

When the camera has a fixed objective (compact/bridge camera, smartphone), the only solution is to cast the image from the telescope with an eyepiece directly on the objective of the camera (Figure 3.1). This is "afocal projection," also called "afocal imaging" or "digiscopy." This is the worst solution, but it helps when we want to exploit consumer imaging equipment. The metering has to be set on "spot" or "center" because the best part of the image is at the center of the field. It is better to set autofocus on, automatic exposure on, and automatic white balance on, but flash off.

The advantages are as follows.

- Operates with any consumer imaging equipment with fixed objective.
- Operates with all telescopes.
- Allows variable magnification and field adaptation by choosing different eyepieces and distances between the eyepiece and the camera.
- Automatic exposure and autofocus correctly operates; the zoom allows a certain degree of freedom.

There are also drawbacks.

- The edges of the image are darkened (due to vignetting).
- The edges of the image are blurred (due to field curvature).

3.1.2 Digiscopy with camcorder, bridge camera, or compact camera

The situations are identical because these devices have a non-removable zoom objective and they normally operate with automatic exposure and autofocus (see how the device reacts in Figure 3.2). Nonetheless, a camcorder has a narrow field of view due to its often greater focal length. A focal-length reducer may be added in front of the lens of a camcorder; this enlarges the field of view and reduces the magnification at the price of an additional vignetting. The zoom brings freedom to focus and adapt the field of view. The best approach is to first try with a long-focal-length eyepiece, for instance 25 mm, then shorter lengths: 15 or 12 mm, then 7 or 8 mm, until the image is satisfying. Digiscopy allows the taking of souvenir

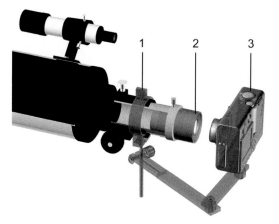

Figure 3.1 A digiscoping adapter (1) tightened on the fixed part of the focuser ensures a perfect alignment of the eyepiece (2) and the camera (3), while not eliminating vibrations when we operate the camera. The same accessory is suited to camcorders.

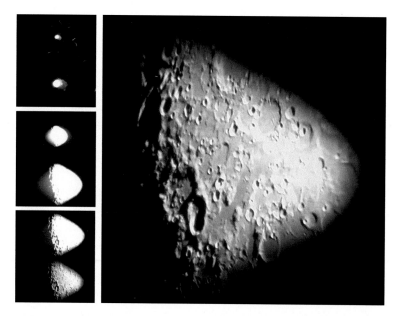

Figure 3.2 An HD camcorder in automatic mode in digiscopy behind a fine apochromatic triplet refractor with an 8-mm eyepiece. Afocal projection produced disappointing results. Left, top to bottom: the images show the Moon as the camcorder is brought closer to the eyepiece. The camcorder automatically adjusts the exposure and the focusing. Right: vignetting, field curvature, and loss of accuracy are obvious despite the use of high-end consumer equipment. The resolution is only 3.5 arcseconds (25° W, 41° S).

Figure 3.3 A compact camera in digiscopy with a spotting scope. The mechanical adapter is screwed onto the M57 thread (around the objective). This avoids leakage of light and guarantees a correct alignment. Image from Wikipedia.

take-home images at star parties, but offers only a very narrow image field unless we use a specific eyepiece.

Specific eyepieces dedicated to digiscopy (starting from $60, up to $380) and mechanical adapters (starting from $40) improve the results.

- Digiscopy-specific eyepieces lower vignetting and field curvature; they may have an M57 thread to directly screw onto the camera.
- Numerous mechanical/optical adapters exist, such as Baader ADPS, Nikon, Sony, Vixen (discontinued, but available on the second-hand market), and manufacturer-specific adapters for birdscopes like Swarowski (DCA, Digidapter …), Leica, Pentax, Kowa … The Panasonic Lumix LX, then GH and GX, along with the Nikon Coolpix families, are some of the most classic choices for digiscoping with birdscopes (Figure 3.3).

Some high-end, integrated adapters (specific eyepiece plus mechanical fitting) are rather expensive, up to $950: this is the price of a telescope and a planetary camera which give better results. Some cheap birdscopes/spotting scopes/binoculars have an in-built camera: tests proved that they provide poor results, but the formula is clever.

3.1.3 Digiscopy with a smartphone

The smartphone is simply held by hand, or with the help of a mechanical adapter, at a distance of 1 or 2 cm behind the eyepiece of the telescope. Since the focal length of a smartphone lens is fixed, focusing has to be set with the help of the focusing knob of the telescope when the smartphone is in place (the focusing is different for the eye and for the smartphone).

Shooting in such conditions is not very convenient because the diurnal motion quickly sweeps the Moon out of the field, while uncontrollable variations in centering, focusing, color balance, and brightness all occur. About 90% of the images might be blurred and badly centered, but a digiscoping adapter (Figure 3.1) for smartphones will help to maintain the alignment behind the eyepiece. As with a compact camera, the best thing to do is try first with a long-focal-length eyepiece, and then with more powerful eyepieces until the field of view is exploited optimally (Figure 3.4).

3.2 Prime-focus imaging

The lens of the camera is directly replaced by the telescope. This requires the imaging device to have a removable objective, as in a DSLR or an industrial camera. This technique is mainly exploited for deep-sky imaging, because it offers the widest field of view and preserves all of the incoming light with minimal magnification. It is suited to imaging lunar phases, earthshine, conjunctions, lunar eclipses, and wide-field subjects. In many cases, short exposures at prime focus do not need a motorized mount. As the signal/noise ratio is maximal, still images are satisfying. On the other hand, if we use large sensors such as full-frame DSLRs, optical flaws at the corners of the image cannot be ignored.

If the telescope has a small focal length, for instance 400 mm, it may be attached to a simple tripod or an unmotorized azimuthal mount. The Moon has to be placed to one side of the image (depending on the hemisphere) to pre-empt the diurnal motion.

3.2.1 Some possible back-focus concerns and solutions

Some optical devices, especially Newtonian (and Dobsonian) telescopes, have their focal plane located close to, or inside, the eyepiece holder, even if the latter is retracted (pushed) to its maximum. To cast the image on the sensor, we must have a small margin at our disposal: the distance between the eyepiece holder and the surface of the sensor, or back focus. In professional telescopes, the back focus may extend for tens of centimeters, even meters (especially the Nasmyth

Figure 3.4 The Moon imaged with a smartphone held by hand behind a 125-mm (5-in) telescope with a 12.5-mm Plössl eyepiece. Note the field curvature at the top right (the image is somewhat blurred), the vignetting (the center is overexposed while the edges are darker), and the appearance of unnatural colors at the center because of the consumer CMOS sensor. The image has been enhanced with simple adjustments of contrast and sharpness. Nonetheless, the result is extremely pleasant for a casual – and actually very first – shooting during a little star party. Image by Mathias Barbarroux with the help of Magali Chaussade.

focus for cumbersome spectrographs). In amateur telescopes, we encounter several cases.

- Refractors have sufficient – even excessive – back focus because they are intended to be exploited with a diagonal.
- Debut Newtonians/Dobsonians have their focal plane inside the drawtube, but serious Newtonians have an enlarged secondary and a focal plane intentionally outside, at about 50 mm for direct DSLR adaptation, while collapsible Dobsonians may have an adjustable distance.
- There are no concerns with catadioptric telescopes, either because their primary mirror shifts along an internal axis while their focal plane is always located outside their fixed focuser, or because they are designed to fit large-focal-length equipment (a big, deep-sky CCD camera with an off-axis guider and filter wheel).

Figure 3.5 Adapters for DSLRs. They have a T2, female thread to be screwed on the focuser (if it has a male, T2 thread), on an additional prime-focus T2 adapter, or on the T2, male thread of a Barlow lens or a tele-extender for higher magnifications. The bottom part is a brand-specific, bayonet mount (here are shown a Canon EOS adapter and a Nikon adapter). M48 adapters have to be directly inserted into the drawtube in place of the eyepiece, with no vignetting, unlike a T2.

3.2.2 Adapting a DSLR to prime focus

Typical DSLRs have their sensors placed at approximatively 45 mm behind the mount. Here are some common examples:

- Nikon: 46.5 mm
- Pentax: 45.46 mm
- Canon: 44 mm

The back focus must be at least equal to the distance to the sensor + T2 or M48 adapter thickness + the distance from the optional eyepiece holder to a T2 or M48 adapter (Figure 3.5). There are several ways to achieve a sufficient back focus.

- Adapt a low-profile focuser (e.g. Lumicon, Antares, JMI ...).
- Use a collapsible truss-tube optical tube (e.g. Skywatcher's Flex-tube and others) or shorten a truss tube.
- Bring the primary mirror cell closer to the secondary, sawing the bottom off a full tube. This has been successfully done on a full-tube, 300-mm Dobsonian, by an experienced craftsman; it requires just a metal saw, a driller, and a certain amount of recklessness.

3.2.3 Adapting a webcam to prime focus

Assuming that the webcam has a standard objective thread of 12 × 0.5, adapters are widely available at telescope dealers (Figure 3.6). The objective has to be unscrewed first. Two solutions exist:

Figure 3.6 Webcam adapters with male, 12 × 0.5 threads to be screwed in place of the webcam lens. They are compatible with most miniature, analog-output video-surveillance cameras. Top: for a T2, male thread. Bottom: to be inserted into the drawtube of a 31.75-mm (1.25-in) focuser. Both have a standard thread for filters.

- T2 adapter, if the eyepiece holder has a T2 male thread or a T2 adapter;
- 1.25-in adapter, for any eyepiece holder.

Unscrewing the webcam objective means removing the appended infrared-blocking filter. This has barely noticeable consequences for lunar imaging (which is not the case for Mars). The very small sensor of a webcam prevents wide-field imaging with a telescope, but it is perfectly suited to long-focus, cata-dioptric telescopes. The image will be undersampled with short-focal-length refractors or Newtonian telescopes. Cheap, small focal-length reducers can enlarge the field while giving at best a 6- or 8-mm-diagonal, correct image.

3.2.4 Adapting a video camera to prime focus

Serious analog- or digital-output video cameras are predominantly equipped with a C or C/S mount. Such objectives were used with 16-mm film movie cameras, and this format is still in use for video surveillance. Miniature,

Figure 3.7 A C-to-31.75 mm (1.25 in) adapter. This one has an internal thread (optics side) onto which to screw a filter.

analog-output video-surveillance cameras are equipped with small objectives with a 12 × 0.5 thread. Consequently, the following possibilities are encountered.

- Using a 12 × 0.5 → 1.25-in adapter (like a webcam).
- Using a 12 × 0.5 → T2 adapter (like a webcam).
- Using a C or C/S → 1.25-in adapter (Figure 3.7).
- Using a C or C/S → T2 adapter.

Mechanical adaptation is not a problem because many telescope dealers provide both adapters. But images may always be overexposed if the camera includes a non-disengageable AGC. If this is the case, we must center the illuminated part of the Moon in such a way that no part of the image shows the limb; therefore the AGC will correctly set the exposure, as for a classic, daytime image.

Another real problem may be, once again, the lack of back focus. An ultimate solution is to unmount the camera and then place the sensor with its printed-circuit board in a 32-mm PVC pipe, externally abraded to fit to the drawtube of the eyepiece holder. This solution has been successfully tested by the author with a miniature video-surveillance camera. The optical axis was not perfectly centered, but this worked.

3.2.5 Adapting a planetary camera to prime focus

Planetary cameras always have removable objectives and they can be equipped with three kinds of mounts:

- C or C/S mount;
- T2 (42 mm) female thread;

Figure 3.8 A T2-to-31.75 mm (1.25 in) adapter.

- 31.75-mm (1.25-in) adapter to be screwed into the C, C/S, or T2 thread (as in Figure 3.8).

Some cameras, like QHYCCD's QHY5 L-II and the Altair Astro GP-Cam, are intentionally narrow enough to be deeply inserted into the eyepiece holder. This is the perfect solution for Newtonian/Dobsonian telescopes that have their focal planes inside the drawtube. With a 1.2-Mpixel, or higher, sensitive sensor, it is now possible to acquire tens of frames with a non-motorized Dobsonian. The image is large enough for one to voluntarily waste several hundreds of pixels in the margins while keeping the central area for stacking.

Most cameras have only a C thread. A C-to-31.75 mm adapter is provided or easily available. Other cameras, such as ZWO's ASI range, have a female T2 thread, a 31.75-mm (1.25-in) adapter, and a C adapter (for a C or C/S lens).

3.3 Imaging with high magnification

This allows dramatic close-ups of craters, rims, lunar domes, and small topographic features. We already assume that the telescope has to be properly prepared, collimated (for a reflector), more or less aligned with respect to the terrestrial polar axis, and accurately aimed at the desired lunar region. But we also have to know what level of magnification is appropriate and how to obtain it by various means. Larger telescopes have better performance, and higher magnifications are possible with no dramatic loss in brightness and contrast, even if the F/D ratio increases. For instance, a Mak with a 90-mm (3.5-in) aperture performs well at F/D = 14 prime focus, while a 180-mm (7-in) one still performs perfectly at F/D = 30.

3.3.1 Variable-or constant-ratio magnification with Barlow lens

One of the most useful optical accessories for high-magnification imaging is the diverging lens invented after Peter Barlow's work in astronomical optics. A Barlow lens is inserted between the focal plane and the imaging device. Numerous instruments for beginners are sold with a poor-quality Barlow lens, resulting in pitiful, blurred, dark images. This led numerous users (including the author) to become suspicious of using it. But modern achromatic or apochromatic Barlow lenses provide excellent images. A Barlow lens suitable for a lifetime's use can be purchased for $80 to $250. It seems that the shortest Barlow lenses may provide a lower image quality because they are primarily optimized for mechanical sturdiness in case they are used with heavy imaging equipment or binoculars. With a light camera, a long Barlow lens is not an obstacle.

Apart from optical quality and sturdiness, two factors must be considered.

- The diameter: 1.25 or 2 in. Imaging with a planetary camera needs a 1.25-in Barlow lens. The Barlow lens is thus lighter and less expensive, though rigid enough to support a 200-g planetary camera.
- The magnification ratio. It must be chosen with respect to the desired sampling rate. Two kinds of Barlow lens are available. In telecentric Barlow lenses, rays leave parallel to the optical axis. This is necessary for some applications, and the magnification rate is constant, for example 2× or 4×. With non-telecentric Barlow lenses (Figure 3.9), the magnification ratio increases with the distance between the output of the Barlow lens and the sensor. Thus, a non-telecentric Barlow lens can have a nominal 2× power, a 2.5× power with a given extension tube or "spacer," or more with a longer spacer. The nominal magnification of a Barlow lens is 1.5×–5×. Sometimes, the nominal magnification is under-rated.

The last point is the interface. Most Barlow lenses have a standard 1.25- or 2-in output and some have, in addition, a male T2 thread, which is exactly the same, simple setup as prime-focus imaging. The input (telescope side) of a Barlow lens generally has a 1.25-in female thread to screw on filters. And Barlow lenses may be piled up to increase the overall power, but, at present, cameras have such tiny pixels that the magnification is strong enough to avoid accumulating mechanical floats and optical flaws.

The choice regarding the power of the Barlow lens depends on the sampling rate (Section 6.1). The Barlow lens modifies the effective focal length as follows:

$$\text{Effective focal length} = \text{Barlow power} \times \text{Focal length of telescope}$$

Focal lengths are expressed in mm.

That is, for instance, with a 4× Barlow lens and a focal length of 750 mm (e.g. a classic, 150/750 Newtonian):

$$\text{Effective focal length} = 4 \times 750 \text{ mm} = 3000 \text{ mm}$$

Figure 3.9 Magnification amounts to casting an image – formed by a given objective/ mirror – onto a larger surface with the help of an additional optical accessory: either an eyepiece or a Barlow lens. The larger the cast image, the lower the brightness per surface unit area. There are two ways to increase the magnification with no loss of brightness: adopting a larger telescope (with a longer focal length but the same F/D ratio) or choosing a more sensitive sensor with smaller photosites. When the image diagonal is magnified twofold, its surface is magnified by a factor of four (i.e. it is squared), while the brightness per surface unit is divided by four (divided by the square).

This is for a telecentric Barlow lens, or a non-telecentric Barlow lens with no additional spacer. Measuring lunar features will allow precise measurement of the actual power of the Barlow lens. Given the desired magnification, we can choose a non-telecentric Barlow lens, which offers the ability to adapt the magnification to turbulence.

A final thing to know about Barlow lenses is the beneficial effect of conical ray paths: optical flaws are somewhat decreased, especially field curvature and coma. Some Barlow lenses act as coma correctors for Newtonians; this is especially useful with a large sensor such as the CMOSIS CMV.

3.3.2 Variable-ratio magnification with eyepiece projection

Eyepiece projection is the alternative solution to the Barlow lens. Of course, we could use both a Barlow lens and an eyepiece simultaneously, but adding the flaws

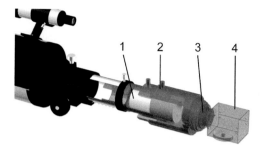

Figure 3.10 Eyepiece projection. (1) Eyepiece. (2) Tele-extender. (3) T2 adapter depending on the imaging device: DSLR (Nikon-T2, Canon-T2, Pentax-T2, and so on), webcam (12 × 0.5 thread-T2 or 12×0.5–1.25-in), video camera (C-T2 or C/S-T2). (4) Imaging device. The distance between the sensor and the eyepiece partially determines the overall magnification according to the simple formula mentioned in the text.

from the two optical elements is not the best thing to do. Eyepiece projection requires a mechanical part called a tele-extender or teleconverter: a simple tube with threads (Figure 3.10).

Eyepiece projection has great advantages.

- Using an eyepiece we already have for observing.
- Having several eyepieces means several possible magnifications.
- Magnification is continuously variable by adjusting the distance to the eyepiece.
- Excellent classic, narrow-field eyepieces are very affordable (e.g. Plössl, Abbe).

Obviously, this solution seems ideal; however, it also has drawbacks.

- Eyepieces are optimized for observing, not imaging.
- Eyepieces may be too large to fit the tele-extender.
- The whole assembly results in a non-negligible leverage effect.
- It may be out of focus with some entry-level refractors intended for observing with a diagonal.

Though apparently simple, a tele-extender (or teleconverter) comprises interfaces, threads, and optional extension or variable-length parts.

- Either a 1.25-in male tube and/or a T2 female thread or a 50-mm (Celestron/ Meade SCT) female thread (focuser drawtube side/SCT visual back side).
- An internal 1.25-in tube for the eyepiece (except for T2-input tele-extenders).
- A T2 male thread output and/or 1.25-in or 50.1-mm (2-in), M48 drawtube (camera side).
- A variable-length part, or extension parts, by means of which to adapt the magnification by varying the distance between the eyepiece and the camera.

Facing these possibilities, the author purchased an SCT tele-extender from Celestron (50 mm SCT "visual back" at the front, T2 at the rear) and a modular tele-extender (T2/1.25 in at the front with a 1.25-in internal tube for the eyepiece, adjustable-length and disposable extension tubes in the middle, T2/1.25 in at the rear) from Baader, covering all needs, as long as the eyepieces are not too large.

Classic, affordable Plössl eyepieces are well suited to eyepiece projection. A large field of view is superfluous if they are used with a small sensor. Furthermore, the dimensions may be critical. For instance, excellent Nagler, Baader Hyperion, and TMB planetary II eyepieces are far too large for most tele-extenders. However, classic Abbe orthoscopic eyepieces perfectly match with planetary imaging: they have only three lenses; they have a very moderate field of view (30°–45°), which is not a problem with small sensors; and they are narrow, rendering them extremely effective for planets and lunar close-ups ($100 for University Optics' phase II Abbe, $150 for Takahashi's Abbe).

The magnification depends on the telescope focal length, the eyepiece focal length, and the distance between the eyepiece and the sensor. This is an approximate formula, because we should consider, for instance, the optical center of the eye lens (observer side) of the eyepiece. In practice, we can use the position of the surface of the eyepiece.

The calculation requires two steps: first we have to know the magnification, then we have to know the effective focal length of the whole optical system.

1 Magnification:

$$\text{Magnification} = (\text{Eyepiece} - \text{sensor distance}/\text{Eyepiece focal length}) - 1$$

Distance and focal length are expressed in millimeters. Here is an example using a 10-mm eyepiece and a distance of 90 mm between eyepiece and sensor:

$$\text{Magnification} = (90 \text{ mm}/10 \text{ mm}) - 1 = 8\times$$

2 Effective focal length:

$$\text{Effective focal length} = \text{Magnification} \times \text{Telescope focal length}$$

Now we can complete the simple calculation, for instance for a telescope with a focal length of 750 mm:

$$\text{Effective focal length} = 8 \times 750 \text{ mm} = 6000 \text{ mm}$$

Eyepiece projection performs best with medium-focal-length eyepieces. Powerful (e.g. 4-mm) eyepieces lead to an excessive effective focal length, whereas long-focal-length eyepieces (e.g. 25 mm) tarnish the image with field curvature. The results are better with a 12-mm eyepiece and a moderate distance between sensor and eyepiece than with a 25-mm eyepiece and a long distance.

The technique is perfect to achieve high magnifications with short-focus telescopes, such as Newtonians and refractors.

3.3.3 Webcam with Barlow lens or eyepiece projection

Once the objective has been loosened, the webcam adapter is put in place. The adapter may have two possible connectors.

- Either a 31.75-mm (1.25-in) tube to be inserted into the eyepiece holder in place of the eyepiece. Thus the webcam can be installed

  - in a Barlow lens or
  - at the 31.75-mm output of a modular tele-extender.

- Or a female, T2 thread for the webcam to be installed

  - at the male, T2 thread of a Barlow lens or
  - at the male, T2 thread output of a tele-extender.

3.3.4 DSLR with Barlow lens or eyepiece projection

The vital element is a female T2 adapter for the DSLR (Figure 3.11). This is a very common accessory, available for major DSLR brands, at a price of $12–40. Some Barlow lenses and tele-extenders already have male, T2 threads that directly match with T2 adapters for DSLRs. Some tele-extenders or Barlow lenses have a 2- or 1.25-in output for eyepieces; in that case a 1.25-in/T2 or 2-in/T2 adapter is necessary, at a price of $35. The assembly has to be stiff because some DSLRs weigh 800 g. Some adapters include both T2 and 1.25-in options, or can be a part of a modular adapter (e.g. Baader). A 12-mm, orthoscopic, Abbe or Plössl eyepiece provides excellent value as equipment with which to start with eyepiece projection (Figure 3.11). A Barlow lens also may be used in place of the eyepiece.

3.3.5 Planetary camera with Barlow lens or eyepiece projection

The system illustrated in Figure 3.11 is also, after a few tweaks, suited to planetary cameras. The tele-extender or the Barlow lens still has a T2 thread or a 31.75-mm

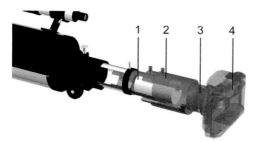

Figure 3.11 A DSLR with eyepiece projection. The eyepiece (1) is inside the tele-extender (2), then a T2 or M48 adapter (3) attaches the camera (4).

Figure 3.12 The whole imaging train: (1) focuser drawtube, (2) removable eyepiece holder, (3) Barlow lens, (4) filter wheel (e.g. luminance, red, infrared-pass filters), (5) filter-wheel drawtube, and (6) planetary camera.

(1.25-in) tube, but the camera has a C or C/S thread, needing a C-to-T2 or a C-to-31.75 mm (1.25 in) adapter.

3.3.6 Assembling the imaging train, focusing, and shooting

In practice, the imaging equipment may be more or less heavy, possibly with mechanical slack and an awkward configuration. We have to securely, but never excessively, tighten the screws, especially if the parts have no brass locking rings to ensure alignment. Figure 3.12 shows a classic configuration for advanced lunar imaging: a Barlow lens, a filter wheel, and a planetary camera.

Prior to shooting the Moon, we orientate the telescope with the help of the viewfinder, assuming that it is parallel to the axis of the telescope, or we can center the Moon with the help of an eyepiece. Then we pull out the eyepiece and replace it with the imaging train. Since there is no reason for the camera to be parfocal with the eyepiece, the image from the camera is now probably out of focus. If we look at the blurred image, some light gradient may appear. We can orientate the telescope toward the brightest area, and then focus.

Then we attach the loose hanging wires (hand controller, camera cable, power supply cable, heating resistor cable . . .) so that no additional vibration may occur while the whole moving part of the telescope can freely track the Moon for the next few hours. Now we can run the acquisition software, with an exposure of a tenth of a second and a strong gain. Even if the Moon is not perfectly centered, its brightness will overflow its apparent diameter; this helps us to center it. A longer exposure increases this overflow in the case of a large offset.

At this moment, the Moon is totally overexposed and unfocused. We can now shorten the exposure and look for the limb or the terminator, because these areas show the best contrast and at least one sharp edge. Focusing may be facilitated by temporarily increasing the gamma parameter (if any) to enhance the contrast. Then we gently adjust the focusing, loosening and then tightening the focusing

105

Figure 3.13 Focusing with the help of stellar diffraction spikes caused by a cross or spider vanes in front of the optics is easy, fast, and very precise. From left to right: the focusing becomes more accurate until the rays fuse. The images have been captured as they appeared in real time on the screen.

knob several times, progressively reducing the travel, until the limb or a large crater at the terminator has been accurately focused. The gain can now be decreased to about 60% (depending on the camera) and the gamma parameter can be reset to a neutral value (the middle of the cursor). Accurate focusing may take 15 minutes if the atmosphere is unstable and if the focuser is not very precise or not properly adjusted in pressure. Focusing at high magnification is far easier with a computer screen rather than on the tiny screen of a DSLR. Here is the most "artistic" part of image acquisition: our eye becomes able to integrate the turbulence by empirically computing a kind of average after some tens of frames.

Another trick is focusing on a nearby bright star (not too far from the Moon to easily re-center it afterwards and to avoid mechanical bending). A cross in front of the aperture, either made of two nylon wires or simply formed by the spider vanes of a Newtonian, causes diffraction spikes from the light of the star (Figure 3.13). When the image is defocused, each spike is double. When the image is perfectly focused, each spike appears as a single, brighter line. Comparable procedures exploit a Bahtinov mask or a Hartmann mask, possibly with the assistance of specialized software.

At this point, the last thing to set is the exposure, with the help of a real-time histogram. Apart from noise resulting from excessive gain and turbulence resulting from atmospheric instability and too long an exposure, there is one other pitfall in lunar imaging: overexposure. This means that bright areas of the image above a certain level are definitely clipped. In Figure 3.14 (these images were taken during different nights), some areas are totally white, overexposed: this cannot be corrected afterwards. Simple visual control during acquisition is not reliable because it depends on screen calibration. The most reliable way to avoid

Figure 3.14 The combination of gain and exposure settings must favor short-duration frames with limited noise. Left: white areas are overexposed; this cannot be fixed at a later stage. Right: correct exposure – and steady atmosphere (10° E, 38° N).

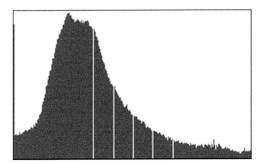

Figure 3.15 In the image-acquisition software or on the screen of the DSLR, the histogram displays in real time the distribution of brightness. If the curve reaches the rightmost part of the diagram, the image will be overexposed.

overexposure is to trust the real-time histogram prior to acquisition (Figure 3.15). Some overexposure comes from quick and unpredictable brightness variations because of quickly evolving turbulence (hence focusing, hence brightness concentration), irregularities in tracking, and gusts. Thus, the targeted area may suddenly show variations in albedo, especially in "young" crater inner rims and steep scarps, because these features are strongly reflective. If these variations are located at the border of the image, they will not have any noticeable impact. All these unpredictable variations are the reason why a margin of 15%–20% below full illumination is safer, and why good acquisition programs display a real-time histogram. DSLR and compact cameras display a histogram in real time or after the image has been taken. If the right part of the histogram meets the right boundary, the image is overexposed, and the values corresponding to the different levels are definitely clipped; no processing is able to fix it. On the other hand, underexposed images are easier to salvage: the areas with the lowest levels of illumination may suffer from noise, but the eye is less troubled by the fact that the relatively dark areas show little detail than by conspicuously "burned" areas of high light level. Stacking helps greatly when we want to bring to light subtle, dark features such as domes and ridges.

3.4 Image-field flaws

3.4.1 Field curvature, coma, and astigmatism

An optical intrument is expected to concentrate light on a plane corresponding to the surface of the sensor. In the real world, the so-called focal plane is a part of a sphere. The goal is to have the sensor dimensions (its diagonal, for instance 1/2 in) matching the size of a part of the sphere close enough to a plane. If this is not the case, that is, when the sphere radius is too short, or the sensor is too large, the image appears curved – and consequently blurred at the edges. Only the circular center of the image is sharp. In most cases, lunar astrophotographers do not experience field curvature because they use small sensors and slow F/D ratios (or large sensors with wide-field optics), so the small field can be considered planar, and the entire image is sharp. Common photolenses have negligible field curvature. With a telescope, the minimal width of the plane field needed for lunar imaging is 0.5°, or 30 arcminutes, which is the diameter of the full Moon. All telescopes have a nearly planar image field exceeding this angular size. But some combinations are very prone to field curvature, resulting in blurred corners.

- Digiscopy is the hardest case. Images are almost always blurred at the corners with common equipment. A smartphone equipped with a wide-field lens is very prone to field curvature, and the only possible fix is increasing the magnification of the telescope with a short-focal-length eyepiece.
- Eyepiece projection with a tele-extender and a long-focal-length eyepiece, for instance more than 20 mm, induces a strong field curvature. The fix simply consists of choosing an eyepiece with a moderate focal length, for instance 12 mm, and reasonably reducing the length of the extension tube to maintain an approximately equal magnification.
- Very-wide-field imaging with a telescope at prime focus and a 24 × 36 DSLR or a camera equipped with a large sensor, such as the CMOSIS CVM4000, not to mention the Kodak KAI-11 000 and other large sensors intended for scientific and deep-sky imaging, is another possibility.

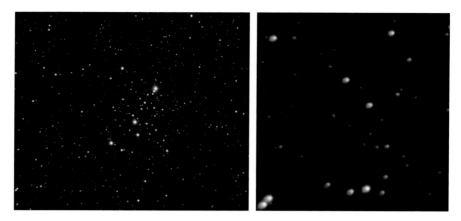

Figure 3.16 Left: the M103 open cluster fills the entire surface of a 1/2-in sensor at the prime focus of a Newtonian. Right: a strong enlargement of the left bottom corner reveals that stars are affected by coma, making them look like the tail of a comet. The center of the full image shows normal, round stars. The cluster is not very bright, and it was not mentioned in Table 3.1. A field corrector diminishes the flaw.

Other flaws are a little more difficult to distinguish because a lunar image is complex. But it is a good idea to evaluate the image field to determine the best optical combination for a given camera. The test procedure is quite simple but it needs a motorized mount. The best way to measure image flaws is to use a grid of points. Given the fact that telescopes often need a distance of more than 20 m (65 ft) to focus, which is generally longer than our hallway at home, we can use stars. Handy targets are open star clusters (Figure 3.16), because they offer numerous, relatively bright stars scattered over wide areas. Table 3.1 presents a short list of bright clusters for both hemispheres.

An exposure of some 2.5–5 s is enough to record stars at prime focus. The image may be color or monochrome. It will almost certainly contain numerous white dots ("hot pixels") and background noise from the camera, but this does not matter because the only relevant information is the shape of the stars, especially at the corners. Note that a semi-apochromatic or achromatic refractor spreads the blue; a red filter improves monochrome images, especially with young stars like those in M45.

By studying various stellar images, we may distinguish, according to the telescope and its configuration (magnification, camera . . .), several flaws.

- Field curvature (Figure 3.17): this results in blurred, but roughly round, stars at the corner, while stars near the center remain sharp.
- Coma: stars at the corners are elongated, and show a kind of blurred tail, orientated toward the edge of the image.

Table 3.1 Bright star clusters may be helpful to test optical flaws in the field of the sensor

Star cluster for field test	Constellation	Coordinates
Pleiades, Messier 45	Taurus	03h47, +24°07
Praesepe/The Beehive, Messier 44	Cancer	08h40, +19°59
NGC 2232	Monoceros	06h26, +04°45
NGC 2451	Puppis	07h45, −37°58
NGC 6231	Scorpius	16h54, −41°48
NGC 2516	Carina	07h58, −60°52

For a narrower field, two prominent globular clusters are good test targets:

Hercules Globular Cluster, Messier 13	Hercules	16h41, +36°27
Omega Centauri, NGC 5139	Centaurus	13h26, −47°28

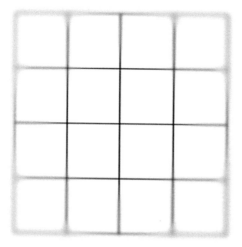

Figure 3.17 This drawing shows the effect of field curvature. This flaw is often prominent with digiscopy, eyepiece projection in certain cases, and prime-focus imaging with large sensors and uncorrected, fast-F/D-ratio telescopes. Many planetary cameras have small, 1/3-in sensors which sit entirely in the planar part of the field: they are not prone to field curvature.

- Astigmatism: stars are cross-shaped when they are best focused. In-focus and out-of-focus images show stars as short, vertically or horizontally orientated lines rather than points. This occurs even at low magnification, with a poor-quality telelens (given the demands of astrophotography).

Note that frames taken during a movie often show offset between one frame and the next, due to gusts and fragile mounts. Since the best result is located where the highest number of frames are superimposed, that is, near the center, the image quality decreases near the borders. This is not to be confused with field flaws.

Recent improvements in catadioptric telescopes – such as the Meade ACF and Celestron EdgeHD – and hybrid formulas – such as Schmidt–Newtonian telescopes – tend to notably enlarge the usable field to match large sensors. Apochromatic refractors often have an optional or built-in field corrector for the same purpose. Newtonian telescopes can be corrected with regard to coma by Ross or Wynne coma correctors. These improvements are useful for wide-field, low-magnification imaging. Whatever the optics, field correctors are very efficient, albeit imperfect, since we have to keep in mind that they are primarily intended for wide-field, deep-sky imaging: stars often show a little loss in sharpness, but the field becomes more homogeneous. Planetary and lunar photography are much less affected by field flaws, thanks to the narrow field related to the high magnifications and small sensors of most planetary cameras.

3.4.2 Vignetting

This classic flaw appears with numerous photolenses and telescopes but it is extremely reduced with specialized, wide-field astrographs (e.g. Takahashi's FSQ range, SkyWatcher's Esprit range, refractors with F/D ratio > 7, Ritchey–Chrétien reflectors). The corners of images are darker than the center (Figure 3.18), due to the optics and obstacles like the drawtube, or lenses that are too narrow (e.g. cheap Barlow lenses with a field lens having a diameter of 21 mm, or even much less than that). This is not a real problem because image

Figure 3.18 Vignetting is the darkening of the corners. This appears when an element is too narrow in the optical path; it is also a drawback of most optical formulas of telescopes. This image (called a "flat field") comes from a 125-mm (5-in) Schmidt-Cassegrain telescope with a 0.63 × focal reducer (increasing vignetting) and the relatively large sensor of an APS-C format DSLR.

editors can easily increase the brightness at the corners, but making a mosaic from images with vignetting is very hard (the figure is asymmetrical and varies). Small sensors of planetary cameras are not prone to vignetting.

To fix vignetting, we have two solutions.

- The first one is inherited from deep-sky imaging. It requires a "flat field," that is, an image (or, better, stacked images) of a white wall or a white fabric lit by a flash. Deep-sky imagers also use lightboxes or flat-field panels. This solution is very efficient, but we will not discuss it because the second solution is much simpler and effective enough for lunar imaging.
- The second one needs an image editor. It consists of a blur selection of the center of the image. Then the selection is inverted, and the curve is set to increase brightness; this easily fixes a reasonable degree of vignetting.

Note: in Photoshop®, the otherwise brilliant function "image → adjustments → shadow/highlight" may introduce rounded, rectangular vignetting, like the "processing → background flatten" functions in Fitswork. This drawback is a normal consequence of these functions; parameters have to be carefully adjusted to avoid artificial vignetting.

To test for vignetting, simply take an image of a bright open cluster in seconds (Figure 3.16) and then strongly amplify the contrast.

3.4.3 Tilt

This purely mechanical flaw blurs images on one side, or in one corner, and arises from a tilted assembly of the imaging device (Figure 3.19). This may result from various situations.

- The eyepiece holder is not aligned (or, for Newtonian telescopes, at right angles) with respect to the light path. The basis of a focuser often comprises three or four screws for precise alignment.
- The drawtube suffers from shifting because the mechanical movements are too rough. A sturdy rack-and-pinion or Crayford focuser is required in order to ensure focusing without axial shifting. This may also be fixed by adding Teflon pads on the drawtube, or, depending on the focuser, by tightening adjustment screws (often small hex/Allen screws sunk into the focuser). Focusing in and out results in inversion of the tilt.
- The sensor is not perfectly at right angles in the camera; this situation is rare but some users (including the author) have experienced it with outdated planetary cameras. A solution is to open the camera – if the guarantee has expired – and add a washer or replace a strut in order to correctly align the printed-circuit board of the sensor.
- There is slack in an adapter. This may happen when an aluminum thread is worn out. The use of brass locking rings strongly diminishes backlash.

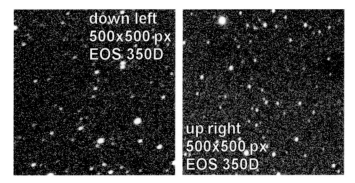

Figure 3.19 Testing tilt by imaging a bright open cluster. Left: at a corner, stars are always blurred and a bit elongated. Right: stars are always sharp at the opposite angle of the image, whatever the rotation of the imaging device, as long as the sensor is large enough to show the flaw. If the flaw came from coma, it might be present at all edges of the image, while tilt is present on one side – or one corner – of the image. Tilt may come from the focuser, the adapter for the camera, the camera itself, or even the objective cell. In this example, it came from a removable part of the focuser. Fortunately, this part was adjustable by means of three screws, and this definitely fixed the problem. Many focusers include three or four screws for this purpose. In another case, the camera was responsible for tilt because the printed-circuit board was not exactly at right angles (rotating the camera showed that the blurred part of the image was always at the same location). A very thin, insulating washer was placed beneath the PCB to correct the angle. Another solution is to add a tilt corrector if the back focus is sufficient.

The diagnosis is easy: if images are always blurred on one side or one corner with various cameras, then the instrument is tilted. If we cannot find the origin of the tilt, a solution is to buy a tilt corrector (costing $80) equipped with three screws to adjust the orthogonality. This adds a distance of about 11 mm (0.43 in) in front of the camera, hence the drawtube has to be pushed into the optical tube by the same length to re-focus, and this may be a concern with small-back-focus Newtonians and some refractors.

3.4.4 Dust

Dust is everywhere in the optics, and this cannot be totally avoided, even if the optical tube is sealed or if it has fans and filters. This has absolutely no impact when dust is located on the objective (or one of the mirrors) because it is strongly defocused, resulting in very slightly darkened areas, with no visible consequences. However, if there is dust close to the sensor, that is, on its protective glass (on the integrated circuit itself, inside the camera), on a removable protection window (often also an infrared-rejection window for color sensors), or at a slightly greater distance, e.g. on a filter or on a Barlow lens, the dust can cause dark rings or dark

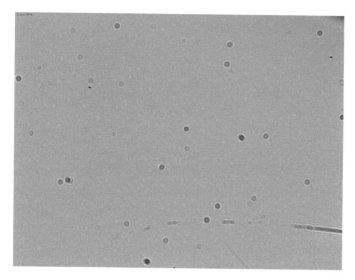

Figure 3.20 This image is a "flat," taken in front of an illuminated surface, after which the contrast has been strongly exaggerated. It shows dark rings caused by slightly defocused grains of dust located on a filter close to the sensor, or on the protective/infrared-rejection window of the camera. The tiny, focused grains of dust are located on the protective glass of the sensor itself. Since the photosites are very small (3.75 μm), the dust is magnified. The horizontal lines are defocused hair. The vertical lines come from the sensor.

circles on the image (Figure 3.20). Furthermore, dust can move while the instrument is being orientated, especially when, facing the meridian, an optical tube on a German-type mount has to be tipped over.

Cleaning small optical parts is rather easy. One needs only glass cleaner and very soft cotton wool, with a cotton bud for peripheral parts of filters, eyepieces, and Barlow lenses. Blowing on the optics is not a good idea, because we could introduce saliva. Brushing is second in the list of bad ideas, because this could scratch optical surfaces. Even though modern optical surfaces are resistant to abrasion, a good technique is to gently wipe the surface with soft cotton and droplets of dye-free, liquid soap, so that the cotton is never dry. Droplets of pharmaceutical ethanol help to remove sticky dust. However, too much liquid could invade the electronics. The cotton has to be replaced after each sweep so as not to scratch the surface with the grains of dust that have been gathered.

Curiously, it is very hard to detect dust before the camera is in place on the telescope. The author has failed to find a convenient method to detect the impact of dust while the camera is not set up on the telescope, and, unfortunately, the game consists of taking a series of test images under operational conditions, removing the camera and cleaning the optical parts, replacing the camera, verifying the quality of images, and so on, until the devices are perfectly clean. Time can

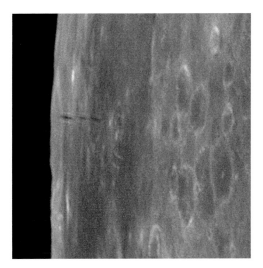

Figure 3.21 The wandering small dark spot on the left is dust. In live imaging, it stays at the same location on the screen while the lunar surface swings or drifts because of turbulence, gusts, irregular tracking, or poor polar alignment. It would be inappropriate to arbitrarily erase dust spots from the final image, because some small feature or other may lie underneath them. But, thanks to inaccuracies in tracking, dust grains often vanish during stacking.

be saved by not tightening the camera between each test image and cleaning, but simply holding the camera in the optical axis of the telescope with one's hand.

When dust is not excessively invasive it can be neglected, as, for instance, when only a few small dark points lie near the borders of the image (Figure 3.21). When dust is present near the center, it can be virtually wiped out with the help of an incorrect polar alignment, or, sometimes, by the effect of moderate gusts, or by irregularities in tracking (the latter never happens with sturdy, high-end mounts). During the acquisition of hundreds of frames, small variations in the orientation of the telescope (a fraction of an arcsecond to some 2–3 arcseconds) cause frames to be shifted. During stacking, a kind of averaged image is calculated by the image-processing software, and the small spots are drowned in the final image: they are spread out and then minimized in intensity because they are considered statistically irrelevant.

3.4.5 Dithering, drizzling, and . . . much too perfect lunar tracking!

In an ideal situation, each detail of the image is exactly cast on each photosite and the final image should be extremely sharp. But the image of the Moon is not composed of square features like the matrix of a sensor. Conspicuous aliasing (Figure 3.22) occurs and the resolution of the final image is disappointing. We

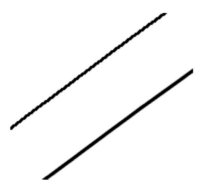

Figure 3.22 Top: the line shows aliasing, like stairs. Bottom: the resolution is better. In practice, a more appropriate sampling rate (with a greater focal length or smaller photosites) or a workaround such as drizzling improves the resolution. Aliasing is conspicuous when imaging at prime focus with large photosites (e.g. a small refractor with a webcam).

must admit that there is a quantization effect: curved features resemble staircases, evoking an old-style computer game with limited resolution.

There are three possible solutions to enhance the accuracy by reducing aliasing.

- Exploiting the "drizzle" function while stacking the images, for instance in RegiStax. This functionality is also provided by some acquisition software, and was developed by NASA/Space Telescope Science Institute.
- Increasing the effective focal length, i.e. increasing the sampling rate (see Section 6.1).
- Adding some irregularities in tracking by manual interventions on the hand controller, or by using deliberately imperfect polar alignment.

These surprising techniques cause variations of position when the image is cast on the sensor. Details can normally disappear because they are exactly at the boundary between neighboring photosites, thus they are partially or totally masked out because the photosites are never strictly adjacent.[1] By introducing small variations of the position of the sensor, these masked details are moved toward the center of the photosites. The idea is to keep in memory all stable details (i.e. those appearing in numerous frames). The larger, stacked image contains more potential resolution, especially at intermediate points of curved or diagonal features. Obviously, the Moon shows numerous rounded and curved features, such as craters and rimae. When the image is undersampled, or when the sensor is not able to acquire vast areas with fine details, using a rough tracking (or a rough polar alignment) and exploiting the "drizzling" function results in a sharper and larger image, at the price of an extended calculation time from the computer.

---

[1] Surface components and various obstacles such as wiring and transfer lines partially mask the photosites, and this is not totally compensated for with the help of microlenses.

Surprisingly, this technique is very efficient in compensating for undersampling. When necessary – for instance, when the image taken in the blue is undersampled relative to that in the near infrared – the resulting sampling is improved when the Hubble Space Telescope Science Institute combines several "dithered" long-exposure images with a deliberate small shifting, and then applies a "drizzling" algorithm to the images.

At this point, we could imagine that the addition of dithering by deliberately imposing tracking drift and then applying drizzling could systematically enhance the accuracy of all images. This has been tried by a number of planetary and lunar astrophotographers, but the results were disappointing. Indeed, this strategy cannot significantly enhance the final image if the original sampling rate was appropriate; the image size is uselessly increased, along with the processing time. Choosing a good sampling rate, e.g. two or three times the expected resolution of the telescope, is effective enough. However, if we want to acquire accurate and large images with a 0.3-megapixel sensor, or if the sensor is not sensitive enough to exploit a strong magnification, the technique is interesting.

3.4.6 Distortion

Distortion misshapes the image in two ways: barrel and pincushion distortion (Figure 3.23). This is not very important if we use a planetary camera, but a 24 × 36 DSLR is more prone to distortion because it has a larger field.

The only possibly noticeable effect of distortion is when we want to realize a mosaic. The boundaries of images are expected to correctly match each other. Distortion can be more or less adequately corrected by an image editor. For instance, with PaintShopPro™, we can use a wide variety of types of distortion in "effects → geometrical effects," or in Photoshop® "filter → distort." This requires a reference image, like a grid, or, at least, a house or a TV antenna shot at various

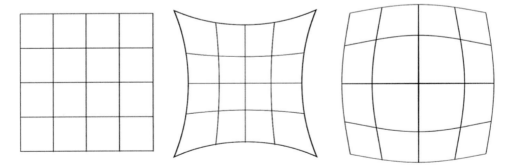

Figure 3.23 Left: normal projection. Center: pincushion distortion. Right: barrel distortion. Image from Wikimedia.

locations within the photographic field. The comparison between a lunar image and a lunar map is not handy because of the librations, but we can rely on the Virtual Moon Atlas. The best solution is using a grid or calibration chart, but this requires a very long hallway or a garden for correct focusing. Fortunately, software for creating mosaics does correct distortion prior to assembling images (Appendix 2).

4

Tuning your telescope for lunar imaging

4.1 Marrying filters, sensors, and telescopes

An experienced lunar astrophotographer knows that there are two ways to obtain good images: having cutting-edge equipment, or cleverly exploiting less-than-perfect equipment. Lunar imaging equipment is a chain, where each link has a role and has to match the performance of the others. We have already seen that using small sensors minimizes the effects of optical flaws, and we guess that excellent optics and mechanics should offer a good starting point. But some tricks may efficiently reduce noticeable flaws, and allow good images to be taken with affordable equipment.

4.2 Fixing chromatism of an achromatic refractor

Since beginners often start with an achromatic refractor, this provides the first opportunity to imagine improvements. When observing or imaging the full Moon, a colored halo appears at the limb and around crater rims because of chromatism (Figure 4.1). By definition, focusing is correct for one color only, or more precisely for a narrow bandwidth centered around a color. Since achromats are designed to match the best sensitivity of the human eye, their best focusing capability is near yellow–green (about 500 nm). In other words, achromats are good monochromatic instruments. By chance, the Moon reflects a wide range of solar light, and it is globally gray, that is, a mix of all colors (we can recall the experiment involving the Newton disk). If we extract the color for which the refractor works best, the resulting monochromatic image is sharp, at the price of a substantial but manageable loss of light. Indeed, the Moon shows differences in color reflectivity because of variations in the composition of the soil, but these differences are greater in the wavelengths approaching the near infrared.

The simplest way to obtain a sharper image from an achromat is to extract the green component of a color image. This is a basic function provided by almost all image editors ("RGB separation"). The red component often has a good accuracy but a low contrast, and the blue component is blurry or completely smeared out by a severe lack of light concentration, whereas the green component is well defined. This is because an achromat is optimized for the human eye (which is most sensitive to green) and color sensors mimic it.

119

Figure 4.1 Top: a color image taken with a DSLR and an achromatic refractor. A purple–blue halo, with traces of yellow, appears near the locations with the greatest contrast: the limb, crater rims, and crater floors. The optics cannot correctly focus all colors at the same time. Bottom: the green layer extracted from the color image confirms that both the achromatic refractor – designed for observing – and the color sensor are intended to work at their best in the green (the human eye is mimicked by a Bayer matrix). A monochrome camera with a green filter (with infrared and ultraviolet rejection) is more efficient (46° E, 12° S).

Chromatism can be lowered also by adding a specific "antichromatism" filter, e.g. the "Chromacorr" or other equivalent filters. Unfortunately, antichromatism filters are often as expensive as the whole achromatic refractor. More recent filters, such as the "fringe-killer" and the "contrast booster," are much more affordable. A "Minus Violet" filter, subtracting light in the violet–blue wavelengths (that is, passing all other colors) greatly improves contrast for observing and imaging by almost canceling the purple halo near areas of high contrast. This is a very affordable solution, which is also common in terrestrial photography, because even the best photolens with a fast F/D ratio (e.g. 1:2.8) has an element of chromatism at full aperture.

An efficient trick is to use a conveniently chosen color filter and a monochrome camera. "Monochrome" means that it records but cannot distinguish colors. As we saw in Section 2.9.7, a monochrome camera has more sensitivity and accuracy than a color camera. Furthermore, it does not show any loss due to a color filter's crossover. This is ideal when used in combination with a high-quality color filter which has steep slopes, that is, an abrupt attenuation of undesired colors (Section 4.4). Consequently, the best images can be obtained with a $12–20 green or yellow–green filter, or a Baader "continuum" filter, or even color gel for stage lighting (prior to purchasing a real, astronomical-quality, color filter), and a monochrome camera.

4.3 Red and near-infrared filters

Surprisingly, even a reflecting telescope – which focuses all colors at the same point with no chromatism –benefits from the use of a color filter. Despite there being a subtle colorimetric differentiation of the components of the lunar soil, lunar light is globally grayish. It comprises all the colors of the spectrum, like a rainbow whose colors are mixed. All colors have a different behavior when they pass through the atmosphere because of their wavelength. In particular, the blue color is extensively scattered (hence the blue sky) and is greatly affected by turbulence. Various phenomena are involved, featuring high-altitude clouds and volcanic dust, air composition, and pollutants, but the most interesting, because we can manage it, is Rayleigh scattering, which expresses how light is scattered by molecules (when the molecules are smaller than the wavelength of the light). The scattering is greater for short wavelengths, depending on the fourth power of the wavelength, hence violet is sixteen times more strongly scattered than red. On the other hand, removing the blue part of the light is equivalent to removing the finest part of the image, like a low-pass filter. In practice, imaging in red light suppresses the details affected by turbulence and scattering. If we think this through a little further, we could imagine that infrared imaging allows better images. Unfortunately, two phenomena occur: water vapor dims the infrared (and the terrestrial atmosphere contains a lot of water vapor, except above dry deserts), and the resolving power is inversely related to the wavelength (Figure 4.2). As infrared has a greater wavelength than red, it is necessary to make a compromise between relative insensitivity to turbulence and an acceptable degree of inaccuracy of the images. A small-aperture telescope (100 mm/4 in or smaller) does not attain a high enough resolution for turbulence to be clearly visible, hence no filter is needed. Large telescopes (e.g. 400 mm/16 in) are very prone to turbulence, but their resolution in the infrared is good enough to match the best resolution that can be achieved by an Earth-based telescope with passive optics. Telescopes of intermediate size can benefit from a red filter: a good compromise is reached between the loss of light and the gain in practical resolution. The loss of light has to be compensated for by a near-infrared-sensitive camera.

Another interesting advantage of a near-infrared filter is its ability to switch off daylight.

4.4 How sensors match filters – or not!

Filters come in two families: classic, tinted filters and dichroic (or interference) filters. Filters of the first category are cheap and efficient, but they show soft slopes,

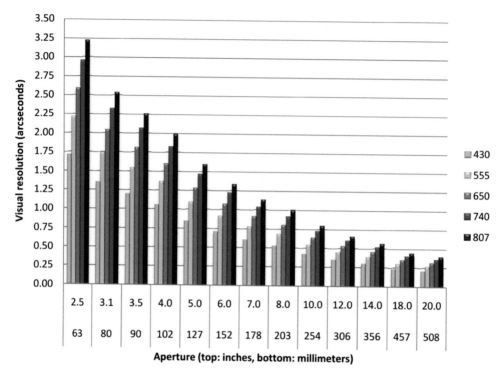

Figure 4.2 This chart shows the theoretical resolution (Rayleigh criterion) of a telescope with respect to its aperture and the wavelength, from blue (430 nm) to the near infrared (807 nm). Imaging easily provides double the resolution. Of course, the use of a near-infrared filter tragically reduces the resolution of a small telescope, beyond the average effect of atmospheric turbulence itself! On very steady nights, a blue filter is best, except for achromatic refractors which barely concentrate short wavelengths. Medium-aperture telescopes take advantage of a red filter, with little loss in light and less sensitivity to turbulence. Large-aperture telescopes have a comfortable light-gathering power, counterbalancing the loss due to the filter, and a large margin of resolution in the infrared.

with leakage of other colors, and their transmission ratio is limited. For instance, a tinted, yellow–green filter (the standard Wratten color code is #11) transmits 78% of incoming light. A dichroic, green filter transmits 96% of incoming light and shows steep slopes: its bandwidth is not polluted by neighboring colors. Another difference is the undesired transmission of infrared and ultraviolet light by classic filters, although the addition of an ultraviolet–infrared-blocking filter is a cheap and very efficient fix; nearly all filters have a thread to stack them. The lower transmission of tinted filters can be rectified by longer exposures, at the price of more sensitivity to turbulence, or by increased gain, at the price of more noise.

The most interesting filters when we wish to minimize turbulence are red and near infrared. Both color and monochrome cameras are sensitive to such wavelengths, but a DSLR is even more filtered than a color webcam because the latter is equipped with a removable infrared-blocking filter. For example, a Canon 10D

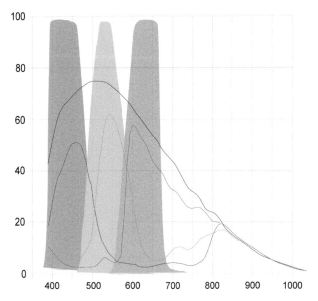

Figure 4.3 The blue, green, and red surfaces show the transmission of dichroic filters. The vertical scale rates the efficiency. The horizontal scale shows the wavelength in nanometers. The colored curves show the sensitivity of a color sensor with its Bayer matrix. The black curve shows the sensitivity of the monochrome version of the sensor. The sensitivity has to match the transmission of the filters. Red or near-infrared imaging needs a good sensitivity in these wavelengths; nonetheless, the exposures are always longer than in visible light. Note the crossover sections where a loss of light is obvious, especially between blue and green. The curves shown here correspond to the properties of on a real sensor.

DSLR has only 7% of its maximal sensitivity at 650 nm. But, even if we remove the infrared-blocking filter, the Bayer matrix is still there. That is why a filter has to match the spectral sensitivity of the camera in order not to waste light. Near-infrared-depleted color cameras (e.g. with Sony's IMX224 or E2V's 76C560 sensors) have at least half their maximal sensitivity near 825 nm, showing a fourth sensitivity peak, in addition to blue, green, and red. Assuming that the cameras are exploited with no infrared-blocking filter, they are perfectly suited to a near-infrared-pass filter such Astronomik's Planet Pro 742 or 807.

For fixing the chromatism of a refractor one basically needs either a minus-violet filter or a yellow–green filter. If, unfortunately, the maximal transmission of a yellow–green filter corresponds to a crossover gap of the Bayer matrix of the color sensor, the loss of light can be significant: the image becomes sharper but very dark. With a color sensor, a #56 green filter is better adapted because its maximal transmission of 79.9% is located at 530 nm, very close to the maximal sensitivity of the color sensor of Figure 4.3. In this case, the simplest solution is the best: a breakdown into fundamental colors in order to keep just the green layer. With the monochrome version, a slight discrepancy appears toward shorter

Figure 4.4 A very sturdy rack-and-pinion focuser for heavy loads. Lunar imaging simply requires a sufficiently good focuser with a load capacity of 200–800 g, ranging from a light webcam to a DSLR or a planetary camera with a Barlow lens and a filter wheel.

wavelength, and the best sensitivity is near 515 nm, corresponding to a #66 filter with its maximum transmission of 84% at 520 nm.

4.5 Tuning the eyepiece holder

Several types exist.

- Rack and pinion, with the advantage of a heavy load capabability, but this kind of mechanism requires precision manufacturing and high-quality materials. The author saw a plastic rack crushed by a brass pinion within three nights. The classic flaw is mechanical slack, causing an irritating default in centering when focusing in and out: the image dances from edge to edge. Nonetheless, excellent rack-and-pinion focusers have been adopted on cutting-edge instruments (e.g. Starlight Instruments – Figure 4.4, JMI . . .).
- Capstan, a very sturdy and precise focuser, but the design has to include the possibility of being able to focus in and out without rotating the camera. It is outmoded, but some manufacturers (e.g. Takahashi) have renewed the design for astrophotography.
- Crayford and reverse Crayford, a very clever design using bearings to eliminate mechanical slack and hard spots. The load capacity is generally weaker than those of the other types, but it is very precise, and some DIY enthusiasts have built their own Crayford focusers.

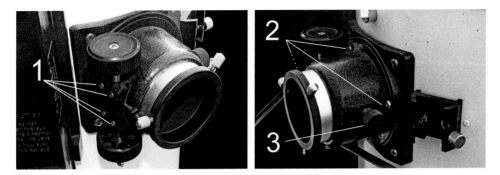

Figure 4.5 Adjusting the firmness (1), adjusting the axis (2), and locking (3) a common rack-and-pinion focuser.

Whatever the technical solution, a serious focuser can be adjusted in three ways (Figure 4.5).

- It can be locked by a knob, when the focusing has been done, to prevent the imaging equipment from sliding down because of its weight, especially when the Moon is high and if the observation site is close to the equator.
- The firmness of a rack-and-pinion focuser can be adjusted with one or four screws under the rack, between the focusing knobs. When we unscrew the cover, we can verify that enough lubricant grease is in place.
- The last optional adjustment is to ensure that the focuser axis is at right angles (Newtonian[1]) or aligned (all other telescopes): this is done by means of three or four screws at the base of the focuser. If the mechanical axis is not precisely aligned with the optical axis, a tilt corrector can be used to bring about a solution with little loss in back focus.

Some focusers can be motorized, or focusing motors can be adapted to a focusing knob. The ASCOM standard (a software layer) brings a convenient solution for remote operation. If the focal equipment remains unchanged for long periods, motorization is valuable because it totally suppresses vibrations when focusing. But a lunar photographer often swaps devices, in order to have a wide variety of magnifications. This requires frequent handling and defocusing with large variations. Even with a variable-speed motor, the action quickly becomes annoying unless the device has a handy disengagement system.

Microfocusers, which have been installed on microscopes for a long time, relatively recently appeared on amateur telescopes. This is a significant step in precision imaging. With a 1:10 reducer, the accuracy in focusing is unrivaled.

---

[1] Some professional telescopes have a third mirror and a Nasmyth or coudé focal plane at right angles to the tube for fixed, heavy loads such as spectroscopes that the telescope cannot support.

Table 4.1 Focusing tolerance in micrometers with respect to the F/D ratio

F/D	4	5	6	7.5	8	9	10	12	15
Focusing tolerance (μm)	32	50	72	113	128	162	200	288	450

Figure 4.6 A disengageable, DIY arm lever. Locking the shaft is done by screwing the wing nut. Unlike with a motorized focuser, vibrations while handling are not suppressed. The accuracy is similar to that of a microfocuser, and this is adaptable to any rack-and-pinion focuser. (1) A bolt maintains the long screw (4) passing through the focusing knob (in place of the original, short screw). (2) Aluminum shaft, about 20 cm long. (3) Wing nut to lock the shaft or to disengage it. (5) Two lockwashers to prevent undesired rotation of the shaft.

A microfocuser is an important mechanical improvement for a Newtonian telescope and any fast-F/D-ratio (F/D < 7) refractor. Catadioptric telescopes often have a built-in precision focuser (needed by design), though some owners of catadioptric telescopes add a classic focuser in addition to the original system.

Another solution is a DIY lever arm, acting as a microfocuser (Figure 4.6). As with a motor, there is a strong need for a disengagement system. The author uses a 20-cm aluminum shaft locked by a wing nut and two lockwashers. Handling is very fast and easy, and the accuracy in focusing is far better than with the original focuser.

The need for accurate focusing is illustrated by Table 4.1. The tolerance for a sharp image is rather narrow with "fast" instruments, because it depends on the square of the F/D ratio. The data in the table have been calculated for 550 nm (green).

A good, old-fashioned achromatic refractor operating at F/D = 12 or 15 is very tolerant: the image is sharp even if the focuser is positioned within an interval of two tenths of a millimeter before or after the precise position of the focal plane. An F/D = 4 Newtonian is ten times as demanding in focusing accuracy.

4.6 Tuning a reflector for lunar imaging

4.6.1 Baffling

Baffling by adding rings inside the tube wall prevents internal light reflections. It may improve contrast and avoid glare when a bright celestial object, such as a luminous star, Jupiter, or the Moon, enters the field. Reflections may come from anywhere in the optical tube, and they are fragments of the unfocused image, scattered by one or more reflections from paint or screws, somewhat altering the contrast, or, in the worst cases (reflecting telescopes), directly illuminating optical surfaces through the focuser prior to the reflection onto the mirrors. Improving (or adding) baffling should not be considered if more important issues, such as the collimation or the quality of the optics, have not yet been fixed.

In catadioptric telescopes, the only possible location for additional light baffles is between the plate and the primary mirror. This is the solution adopted by some serious manufacturers such as Lomo, Intes Micro, and others. Many observers have reported an excellent contrast and an obvious lack of light reflections with such optics. The rings have equal internal diameters because the first part of the light path is cylindrical.

Refractors represent the simplest case: as they consist of a simple, straight tube, a series of internal rings, with a decreasing free internal diameter from the lens to the focuser, can be placed along the light path. Simple refractors have at least one ring, whereas cutting-edge refractors have multiple rings. If we want to add rings in order to acquire a better contrast, they simply have to be glued along two or three shafts, and then slid into the tube. If the original ring cannot be removed prior to installing the series of rings, we just have to unscrew the focuser to access the ring, and then firmly drag it outside the tube. If it has been glued into place by the manufacturer, we can build two sets of rings with their shafts, one to be placed between the previously loosened lens cell and the fixed ring, and the second between the fixed ring and the loosened focuser. An objective lens forms a light cone, hence the successive rings have to offer increasingly narrow holes the closer they are to the focuser. A very good illustration is given by Dick Parker (http://mirrorworkshop.mtbparker.com/refractorStory.html) for his beautiful home-made refractor. Newtonians also have a non-cylindrical light path. The classic book *Telescope Optics* by Rutten and van Venrooij explains Newtonian baffling in detail. The simplest way to baffle a Newtonian telescope is to add a ring just above the primary mirror, masking the edge of the primary by some 5–7 mm (0.24 in). This idea is presented by H. R. Suiter (http://home.digitalexp .com/~suiterhr/TM/Topten.htm#01), taking into account the fact that the edge of the primary mirror is its worst part in terms of curvature. It acts both as a single baffling ring and as an "iris" which may globally improve the diffraction figure at the price of a small decrease in light transmission and resolution (fortunately, a

Newtonian telescope has plenty of transmission and resolution by virtue of its large aperture). If we want to add whole series of internal rings to a Newtonian telescope, an excellent method is provided by the on-line "Newt-on-the-web" software (http://stellafane.org/tm/newt-web/newt-web.html) by Kenneth H. Slater. This software accurately computes the properties of a baffled Newtonian telescope. Despite the apparent complexity of baffling a Newtonian because this requires unevenly located rings, each with a different internal diameter, the optical tube is large and fully accessible. The baffling rings can be made of cardboard, PVC, MDF, plywood, or aluminum. The surface should not be smoothly ground because a certain amount of roughness scatters light, hence dimming reflections. The internal edge should resemble a razor blade to minimize the reflective surface, viewed from the open side of the tube. The paint must be matte black. The best way to attach the rings is to glue them to the shafts with epoxy resin or a cyanoacrylate-based glue. The otherwise reflective droplets of glue must also be painted matte black. The structure (rings and shafts) can be put in place and maintained by frictional forces until the solution has been proven to offer an improvement in contrast and merits fixing permanently with screws.

Finally, baffling is only one component of a good telescope. Its absence does not mean that the telescope has poor optics or mechanics, but, on the other hand, some of the best telescopes (Rumak and apochromatic triplet) that the author has used did have a substantial baffling. It is important to notice that these telescopes have air-current evacuation, because in open, unbaffled tubes, hot air rises and then flows away, while closed and baffled tubes keep hot air inside and at the top of each baffling ring. To remedy this, some tiny holes may be prepared along the edge of each ring. In other cases, the tube itself has holes. The most difficult case to manage is that of carbon-fibre refractors with numerous baffling rings. Since carbon is an excellent insulator, heat cannot escape from a sealed tube. Some large, high-end refractors, such as the Officina Stellare and others, have evacuation holes and show no internal heat accumulation despite the presence of numerous baffling rings.

4.6.2 Painting and flocking

A really, absolute matte black paint helps to dim light reflections considerably. Numerous expensive and large-aperture telescopes have too much reflective paint, and this has led amateur astronomers to re-paint the internal tube walls with spray paint for cars. A certain roughness of the paint layers has to be achieved in order to scatter reflections (this is why the rough surface of the Moon is not so reflective). Some very fine grains (sawdust, fine sand, crushed sugar) may be placed on the walls prior to painting, in order to obtain a slight granularity.

Some trials must be performed on small surfaces to ensure that the spray paint is less reflective than the original paint. Generally the spray gun must be placed about 25 cm (10 in) from the surface and used with brief applications. The very

Figure 4.7 A simple paint cabin assembled with two trestles, two planks, and a tarpaulin. The disassembled, naked optical tube is totally covered by the tarpaulin. The outside of the tube was painted first. It lay on a broomstick to facilitate its rotation without having it touch the ground.

first bursts often project unwanted droplets, so it is safer to begin each painting session on a trial surface. If the result is satisfying, the next step is to build a paint cabin. For this, one requires two trestles, two planks, and a large tarpaulin for the construction site. When all of the optical and mechanical parts of the optical tube have been removed, the naked tube is externally covered by a canvas sheet (Figure 4.7). The tube can now lie in the cabin, ready to be painted with very fine and brief applications of spraying, avoiding the formation of droplets, without superimposing layers and without painting the vertical walls. After a first 15-cm (6-in) wide band has been completed and before it has dried, the tube is rotated by a fraction of a turn to paint a second band, and so on until the first paint layer has been completed (Figure 4.8). The second layer should be applied on a subsequent day.

Another (or complementary) trick is placing adhesive black velvet in strategic places. General consumer adhesive velvet and specialized flocking material have been used. The goal is to dim and scatter light reflections, as with baffling, but flocking is applied especially to the surfaces at right angles to the focuser (Figure 4.9). This is commonly done with Dobsonian and truss-tube Newtonian telescopes. The material must be perfectly cleaned with a vacuum cleaner, otherwise dust particles will be deposited on the primary mirror. Small surfaces of material are needed: about 30 cm in height and width, in front of the focuser. Flocking material should not be applied close to the primary mirror because it could retain dust and moisture. For the same reason, and because they are

Figure 4.8 The bare tube is externally covered by a protective sheet. The spray paint is applied in successive bands of width about 15 cm (6 in) along the whole length of the tube. The tube is rotated by a fraction of a turn between each application of a band of paint and the next. Some three coats are needed, with one-day intervals between application of each layer and the next.

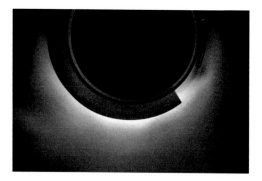

Figure 4.9 Ordinary flocking material – adhesive black velvet – is glued in front of the focuser (after a serious vacuum-cleaning session to remove dust and cat hairs). This significantly dims light reflections, but cannot be used all over the internal walls because the material could retain dust and moisture. Note the difference in reflectivity between the flocking material and the matte black paint.

normally closed, catadioptric telescopes are not often flocked (matte black paint and baffling are more convenient). Refractors are not flocked as long as the internal paint is dark enough, in addition to the effect of baffling rings, if they are present.

Here are some suppliers of flocking material (the list is not comprehensive):

www.scopestuff.com/ss_flok.htm
www.fpi-protostar.com/flock.htm
www.firstlightoptics.com/misc/black-velour-telescope-flocking-material.html

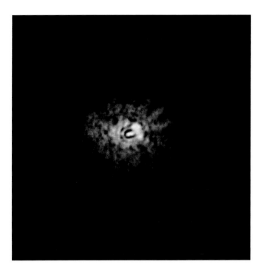

Figure 4.10 Image of a focused star at high magnification. The diffraction pattern is strongly altered by simultaneous optical flaws and improper adjustment of the optics. The collimation was optimized.

4.6.3 Mirror stress, astigmatism, and offset of a Newtonian telescope

Measuring all optical flaws requires professional equipment, and amateur astronomers are rarely qualified opticians (though some indeed are). But some flaws can be detected by observing a star at very high magnification. One of the author's Newtonian telescopes showed oval stars and rough diffraction patterns as illustrated on Figure 4.10. It simply needed some adjustments.

To offset or not to offset

Before deprecating the optical quality, we can take a first look at the secondary mirror. Owing to the shape of the light path in a Newtonian telescope, the secondary mirror has to be moved slightly away from the focuser. As the secondary is elliptical, its lower edge, closer to the primary mirror, is inside the light cone from the primary, while its upper edge, closer to the eyepiece holder, is outside. This optical misalignment leads to a loss of illumination of the secondary, so that the magnitude limit is lowered. This is significant for *deep-sky* imaging. For *lunar* imaging, the most important thing is to obtain a symmetrical diffraction pattern, no matter whether the secondary mirror wastes a substantial proportion of the illumination from the primary mirror. The basic formula is

$$\text{Offset} = \text{Secondary small axis}/(4 \times \text{F/D ratio})$$

where the offset and secondary small axis are expressed either in millimeters or in inches.

Here's an example with a 58-mm secondary mirror and an F/D ratio of 4.7:

$$58 \text{ mm}/(4 \times 4.7) = 3 \text{ mm}$$

In this Newtonian telescope, the center of the secondary was originally shifted by 7 mm. Despite its good-looking sketch, the secondary was surprisingly incorrectly centered. The secondary was glued with double-sided, adhesive foam, a good solution to evenly distribute the forces applied to the mirror. Cutting the foam with extreme caution, after placing a piece of fabric under the mirror (in case it fell), and then re-centering it conveniently solved a major part of the problem, since the diffraction pattern of a focused star appeared much more circular. But, in this case, and because the telescope was devoted to *planetary and lunar imaging* (this case may be of interest also to observers of double stars), the secondary was re-centered, with no offset, to obtain a symmetrical diffraction pattern from a symmetrical obstruction.

Astigmatism

For safety during transportation, the main mirror is often overtightened on its cell. This slightly stresses the mirror, leading to a kind of astigmatism. Detecting it is easy: on de-focusing a star at high magnification, the inside-focus image shows a vertically elongated image and the outside-focus image shows a horizontally elongated image, or vice versa. The star cannot be round or a pinpoint, but instead resembles a short line. Main-mirror stress can be fixed by gently and evenly loosening the locking clips by an eighth of a turn (Figure 4.11). The mirror should turn almost freely on the cell when handled, but should remain stable, and, of

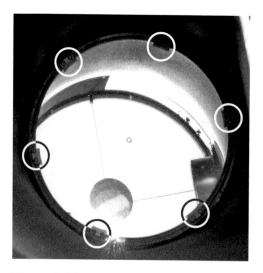

Figure 4.11 A part of the main mirror stress may be decreased by gently and safely loosening the clips.

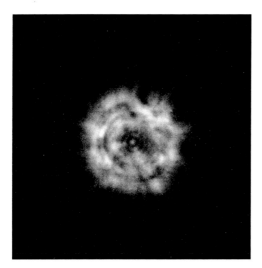

Figure 4.12 Diffraction pattern obtained with the secondary mirror offset set to zero, rotation of the primary mirror by 120°, and slight loosening of the clips of the primary mirror. The image is more symmetrical, including in- and out-of-focus situations, though remaining flaws (optical surface roughness, the shadow of the drawtube, masking by clips) still affect the pattern.

course, it must never be able to fall down when the telescope is aimed at the horizon.

Re-centering the secondary mirror and slightly loosening the main mirror did not suffice, since the astigmatism, while reduced, remained conspicuous. Taking advantage of unmounting the main mirror to clean it, rotation of the main mirror by a third of a turn lowered the astigmatism, with the help of a random (albeit expected) counterbalancing of the optical flaws of the two mirrors. This operation was incompletely but undoubtedly successful, and the diffraction images were much improved (Figure 4.12).

4.6.4 Spider vanes, secondary-mirror diameter

Some telescopes are equipped with curved spiders (e.g. some Dobsonians, TAL Klevtsovs, fine but discontinued Arcane and early Orion Optics UK Newtonians). The idea is to distribute diffraction from the spider vanes all over the image, rather than concentrating it in spikes. Spikes from classic, straight-vane spiders (Figure 4.13) form crosses around some bright stars in deep-sky observing, and around numerous stars in deep-sky, long-exposure imaging. Three-vane spiders show six spikes, four-vane spiders show four spikes, and single-vane spiders show two spikes. The length depends on the brightness of the star and on the exposure duration. The only noticeable concern is the thickness of the vanes, which

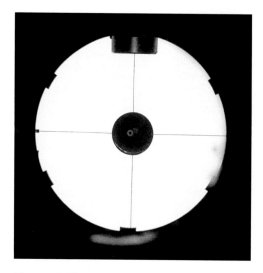

Figure 4.13 Newtonian telescopes are very good lunar imagers, but they suffer from obstacles in the light path: secondary mirror (center), spider vanes (cross), drawtube of the focuser (top), and six (or four) clips keeping the main mirror in place.

enlarges the spikes and really does affect images. Most modern telescopes have vanes of thickness 0.5 mm. Curved vanes produce a little more diffraction than straight vanes, but, in both cases, spikes hardly affect lunar imaging as long as the vanes are narrow, that is 0.5 mm (0.022 in) for up to 380-mm (15-in) telescopes, or 0.7–1.0 mm (0.029 in) for 560-mm (22-in) telescopes, assuming that these telescopes are equipped with a small secondary mirror corresponding to a moderate F/D ratio of 5 or more. Curved vanes represent only the simplest example of apodization.[2] In practice, even vanes representing in thickness up to 5% of the diameter of the main mirror have a negligible effect on contrast for planetary imaging, whereas the diffraction pattern is not as fine as with very narrow vanes, being straight or curved.

The diameter of the secondary mirror brings another obstruction to the light path; it dims light and causes diffraction. Planetary and lunar telescopes are expected to have the smallest secondary mirror possible. In terms of relative diameters, the secondary mirror should ideally represent 20% of the primary mirror for the image to be apparently unaffected by obstruction. But this rule is for *visual use*, not imaging. Stacking and processing of images will lower the effects of obstruction, especially for contrast. For instance, astrophotographers shoot the Moon with Schmidt–Cassegrain and Ritchey–Chrétien telescopes with an impressive obstruction of 33%–40%, and their images are quite sharp and contrasted. Newtonian telescopes and variants like Maksutov–Newtonian telescopes

[2] See www.ejournal.unam.mx/rmf/no511/RMF51116.pdf.

may have an obstruction of 25% or less, and, of course, refractors (and some peculiar telescopes such as a Shiefspiegler or a Newtonian with an off-axis secondary) have no obstruction at all.

Adapting a larger or smaller secondary is possible, if you are prepared to have a telescope that is designed for a particular purpose, such as deep-sky imaging or planetary observing. Most telescopes are multi-purpose, used for everything, but excelling at nothing. A good image stands only in the fully illuminated field, with very few flaws (coma, field curvature, vignetting, distortion . . .). The width of the fully illuminated field sticks to the optical formula, and this creates a challenge for telescope designers who want to offer the widest correct field to deep-sky imagers with large sensors. Fortunately, lunar imaging often needs only a small field with high magnification. We can calculate the *minimal* secondary diameter (we indicate the minor axis, with all dimensions either in millimeters or in inches) with the help of simplified formulas.

For a Cassegrain telescope:
Minor axis = Primary mirror diameter × Distance between secondary mirror and focal plane/primary mirror focal length
For a Newtonian telescope:
Minor axis = Distance between secondary mirror and focal plane/(F/D ratio)

In practice, secondary mirrors are slightly larger than the minimum, in order to entirely receive the fully illuminated field, and because some secondaries are secured by clips, masking out a part of the field. For instance, the calculation indicates 54 mm (2.1 in) for a 254-mm (10-in), F/D = 4.7 Newtonian, but the actual secondary minor axis is 58 mm (2.3-in).

With a Newtonian telescope, the distance between the focal plane and the secondary mirror can be measured by attaching a white projection screen (some inches across) perpendicularly to the focuser, and then casting the Moon onto it. Projection onto a screen made of glossy paper glued on cardboard ensures that the image of the Moon is sharp, and hence perfectly focused to optical infinity. Maintaining the exact position of the focuser, with the optical tube aiming at the zenith, we can put a ruler above the tube and measure the exact distance, with the help of two plumblines sliding on the ruler, with one placed above the secondary mirror, and the other placed just above the surface of the projection screen.

With a Maksutov/Rumak/Schmidt–Cassegrain telescope, the measurement cannot be realized without knowing the precise thickness of the meniscus and (except for pure Maksutovs) the thickness of the secondary mirror with its assembly. We must not touch the secondary mirror with a ruler. A possible procedure starts as above, by casting the Moon onto a sliding screen, with the screen being kept in place and attached to the focuser after precise focusing. Then we can use an indoor mini-laser rangefinder at the exact place of the projection screen. Cheap rangefinders exist, starting at $50, measuring a minimal distance of 5 cm (2 in) with a precision of 1.5 mm (0.06 in) or much better. Since secondary mirrors of catadioptric and Cassegrain telescopes are not planar, any slight angular error when aiming the rangefinder will result in reflections of the beam. So we have to be very careful about where the device is pointing, because an astronomical telescope mirror is not a common target for consumer rangefinders.

Detailed information about telescope optics can be retrieved on various websites, especially the following ones:

www.telescope-optics.net/ by Vladimir Sacek;
http://stellafane.org/tm/newt-web/newt-web.html, collective, from the famous Stellafane convention;
www.atmos-software.it/Atmos.html by Massimo Riccardi, software and tips for telescope design;
www.astrophoto.fr/obstruction.html by Thierry Legault;
www.damianpeach.com/simulation.htm by Damian Peach;
http://freeware.intrastar.net/telescope.htm by Alan Sawicki, a list of software for telescope design;
www.lambdares.com/ commercial (with educational version) software for tele-scope design;
www.myoptics.at/modas/ by Ivan Krastev, freeware for telescope design;
http://aberrator.astronomy.net/index.html by Cor Berrevoets, star-testing simu-lation, including obstruction.

And some reference books:

- *Telescope Optics – A Comprehensive Manual for Amateur Astronomers*, by Harrie G. J. Rutten and Martin A. M. van Venrooij (Willmann-Bell), an excellent book for advanced amateurs.
- *How to Make a Telescope*, by Jean Texereau (Willmann-Bell), an older, world-wide reference, expanded and re-printed.

4.6.5 Re-polishing or replacing the mirror

Assuming the telescope is correctly collimated, the last important optical flaw is surface roughness. This cannot be fixed without testing, de-aluminizing, grinding, polishing, re-testing, and re-aluminizing the mirror. A classic Foucault test is a simple and effective way to assess the shape; a Bath-interferometer test is more impartial (both can be realized in an astronomy club). The contrast can also be estimated with a Ronchi test. Some industrial manufacturers provide good optics with a complete optical test in two or three colors (errors are less prominent in the red). On the other hand, a 10-in mirror can be greatly improved, by some companies and craftsmen, at a cost starting from $500. The only concern is safe transportation of the mirror in a sturdy, wooden housing with adequate internal protection. The expense is reasonable because a good mirror provides beautiful images for decades.

Here is a non-comprehensive list of some companies and craftsmen able to refigure a mirror or to tailor a mirror with a specified optical quality.

USA:

www.discoverytelescope.com
www.galaxyoptics.com
www.pegasusoptics.com
www.lightholderoptics.com/index.htm
www.loptics.com
http://opticaleds.com/
www.opticwavelabs.com
www.ostahowskioptics.com
www.swayzeoptical.com
http://zambutomirrors.com/index.html

UK:

www.nicholoptical.co.uk/index.htm
www.oldham-optical.co.uk/index.htm
www.orionoptics.co.uk

France:

www.mirro-sphere.com/topic/index.html

Germany:

www.alluna-optics.com (mirrors over 300 mm only)

Italy:

www.reginato.it/index.html

Australia:

www.telescopes-astronomy.com.au

Local astronomy clubs also may be equipped for mirror testing, grinding, and polishing.

4.6.6 Getting rid of the secondary mirror?

Some deep-sky astrophotographers remove the secondary mirror or build their telescope with no secondary. These instruments are equipped with a CCD camera at prime focus, with a focusing device (often a classic focuser held by vanes). Such telescopes are intended for pure astrophotography, and substantial obstruction from the camera is not an obstacle. Of course, like some convertible Cassegrain/ Newtonian/Nasmyth telescopes (with disposable mirrors), they cannot be instantly swapped from one configuration to another: the operation takes a couple of hours or more, including the collimation. But many amateur astronomers own two or more telescopes, of which one instrument is devoted to observing while others are reserved for astrophotography.

There are noticeable advantages in replacing the secondary mirror by a camera:

- no collimation of the secondary
- no dew on the secondary
- fewer mechanical parts and less weight
- reduced optical flaws (one mirror only)
- more reflectivity.

This solution matches telescopes having an aperture of 200 mm (8 in) or more, because the obstruction from the camera should not exceed the obstruction of the secondary. With extremely narrow cameras (e.g. QHY5 L-II/III or equivalent products like Orion's All-In-One, the Altair Asqtro GP Cam, and others), the obstruction may even be strongly reduced because they are intentionally designed to have a diameter of 31.75 mm (1.25 in). This is extremely interesting because it is not larger than the diameter of the secondary mirror of a 150-mm (6-in) Newtonian, and the small relative diameter is even more advantageous for larger telescopes.

The classic degree of obstruction is

- 40% for a Ritchey–Chrétien (F/D = 8)
- 33% for a Schmidt–Cassegrain (F/D = 10)
- 28% for a Newtonian (F/D = 5).

This means that a 1.25-in camera perfectly fits a 6 in, F/D = 5 Newtonian. A 60-mm-wide camera matches a 10 in, F/D = 5 Newtonian. It is even better if the camera is smaller than the secondary. Of course, the camera needs accessories:

- a focuser;
- possibly a removable filter;
- a Barlow lens or a tele-extender with an eyepiece;
- wires (e.g. a USB cable, possibly a filter-wheel cable, maybe a power supply for active cooling of some planetary cameras).

A planetary camera produces heat, especially when it is cooled by a Peltier module (this type of cooling is also called TEC, for thermo-electric cooling). The stream of hot air is expelled by the fan, and this creates a turbulent flow through the light path. Even if the camera is not electrically cooled, the housing is deliberately designed to act as a radiator, especially for high-FPS cameras.

On the other hand, as of 2016, no planetary astrophotographer has definitively proved that a "secondaryless telescope" performs better than a classic one. The main advantages to be expected are the simplification of collimation and a non-negligible gain in mass, the latter meaning fewer vibrations. The only commercial products with no secondary mirror are deep-sky imagers (e.g. Fastar and

Hyperstar). A few DIY enthusiasts have built or adapted Newtonian telescopes without secondary mirror (for instance Arthur L. Whipple, see page 7 of http://christian.viladrich.perso.neuf.fr/astro/instrument/solaire-HR.pdf, in French, and Jan Fremerey, www.astro-vr.de/100901_3694.jpg).

At this point, one might think that a decently performing camera and a stable adaptation (employing a Barlow lens or other accessories) could be permanently attached by vanes. The use of a filter wheel, or a frequent permutation of filters, is clearly impractical, but the Moon is mainly monochrome, and, if we want to capture its subtle colors, good color cameras exist. The focuser is the only real problem. In large Newtonians, a standard focuser for refractors, possibly motorized, like an ASCOM-driven focuser, can relatively easily be adapted on a strong spider. A simple and cheap solution is to add a threaded rod sideways to push and pull a single vane in front of the open side of the telescope, with the camera attached to a single vane (this approach was inspired by the old Vixen 150/750 and 130/720 Newtonians).

4.7 Preparing a telescope for lunar imaging

When the main mirror is larger than 200 mm (8 in) and rather thin, it tends to deform under its own weight. One night, the author was observing to the west, then the Moon disappeared at the horizon while Saturn was rising to the east. The 300-mm Newtonian with a light mirror gave a horribly blurred image. The mirror retrieved its shape within 15 minutes, whereupon the image was sharp again; this was not due to the turbulence above the horizon, because the image was steady, albeit blurred. What had happened was the reversal of the optical tube from West to East, on passing the meridian. In other words, the mirror was flipped. Conic mirrors (e.g. Royce's, Schmidt–Cassegrains, and some Dobsonians) are less prone to deformation. However, a little compression on attachment clips (some Dobsonian mirrors are held by a strap) may happen. The slight shifting of a Schmidt–Cassegrain telescope alters only the focusing and centering, not the shape of the mirror. Moreover, azimuthal mounts do not imply reversal of the optical tube, whereas German-type mounts do. This is why we strongly advise that the tube be allowed to cool down in the approximate position planned for the imaging session, corresponding to the future position of the Moon one hour later (this cooling may take two hours for large-aperture or Maksutov telescopes). Hence, the mirror will regain its normal shape while cooling at the same time. No deformation has been reported with refractors, even with heavy 150-mm (6-in) triplet objectives.

A convenient method for preparing a large (more than 200-mm/8-in) telescope is adopting a precise order for the following steps:

- polar alignment and then assembly of the telescope
- orientation of the instrument roughly toward the planned location of the Moon one or two hours later

- cooling down
- collimating.

If the night is humid and the telescope is orientated toward the zenith, the mirror (or the meniscus, or the corrector plate) has to be protected but not covered, in order to let cold air reach the optics.

A mirror with a diameter of less than 200 mm (8 in) is less prone to deformation. Cooling down is the main measure: the optical tube is simply put on the ground, isolated from damp grass by cardboard or a tarpaulin.

The steps for an intermediate (up to 200-mm/8-in) mirror are as follows:

- cooling down
- polar alignment and then assembly of the whole telescope
- collimating.

If the weather turns bad, then just the cooling optical tube must be brought back home, without disassembling the whole telescope. This is a good way to always be ready with little effort. This is one of the reasons why small telescopes are used more often than big ones. Keeping the telescope ready to use in a vented, small observatory improves its exploitability.

4.8 Polar alignment (for lunar imaging)

To correctly track the Moon, the mount need not be perfectly aligned on the rotation axis of the Earth. Indeed, an imperfect alignment improves lunar images because

- it brings about a drizzling effect, decreasing aliasing in the case of undersampling;
- it shifts the position of dust particles on the sensor relative to the cast image of the Moon, wiping the dust from the image;
- the field rotation is negligible, and will be annulled by the registration and stacking process.

Firstly, the mount has to be correctly leveled with the help of a spirit level. Secondly, we have to know the date and time in Universal Time (UT) and the location of the site in latitude and longitude. This is required in order to inform the GOTO system and/or properly orientate an equatorial mount with a polar finder. If the mount is a German-type one and lacks a polar finder, then, after the *rough method* mentioned later, the optical tube could be orientated at 90° with respect to the celestial equator and Polaris should stand in the field of view of a long-focal-length eyepiece. The southern hemisphere lacks a conveniently placed polar marker.

For an equatorial mount (German-type or a fork mount with an equatorial wedge) the following statements apply.

- The ***rough method*** consists of horizontally aligning the basis of the mount (pan or azimuth) with the help of a compass while the polar axis is tilted according to the latitude of the site. For this one does not require a polar finder.
- A better method is using a polar finder. Many kinds exist, and we have to refer to the user's manual. Skywatcher-type and Losmandy-type (Kenko) finders show Ursa Major and Cassiopeia with the locations of Polaris (North), Achernar, Octans, and Crux (South); since the engraved drawing corresponds to a minia-ture version of the optical field, the stars cannot be superimposed onto the markers, which merely indicate the directions of the constellations. Vixen-type finders show the positions of Polaris and Octans, and the date and time engraved on the rotating circle must correspond to an index. Several variants exist, even for the same brand, depending on the year of manufacture. The position of Polaris slowly changes with time, but its variation is neglectible for lunar imaging.
- Optional polar finders are available for fork mounts with an equatorial wedge; they are to be attached at one side of the fork. They are similar to polar finders for German-type mounts.
- King and Bigourdan methods are rather time consuming (taking no less than half an hour), but very accurate and intended for deep-sky imaging.
- Takahashi and IOptron polar finders accurately indicate the locations of Polaris and Sigma Octantis with respect to their drifting over the years.
- Travel mounts also have polar finders indicating constellations both for the northern and for the southern hemisphere.
- If the polar finder is not illuminated, a red flashlight placed almost in front of the finder is very helpful as a means to illuminate the engraved reticle.

For an altazimuthal mount the following criterion applies.

- Initializing the GOTO is sufficient for the embedded software to track the Moon.

4.9 Balancing the telescope

The whole telescope, including the imaging equipment, has to be correctly balanced in all configurations. The purpose is to minimize the imposition of stress on the motors, because they have a limited torque despite their speed-reduction gear, irrespective of whether they are direct-current motors or stepper motors. In addition, some drives have polyamide gears, and even brass gears are not extre-mely stiff (the best gears are made of bronze). There is a second, obvious reason why the telescope has to be balanced: wind.

Equatorial fork mounts may be somewhat hard to balance. When the telescope is aimed at the meridian (where the Moon is at its highest point during the night), the mount seems to be balanced. However, if we orientate the telescope to the East or to the West, the optics will tend to fall down. With small telescopes (up to

150 mm/6 in) this is not an issue, but for heavy optics one requires a counterweight sliding along a shaft screwed into the tube. Numerous Schmidt–Cassegrains have such threads ready for options such as a piggy-back attachment or an additional finder. This is only for longitudinal balancing, but it is often adequate because the additional rear mass compensates for the destabilization of the front side. Other fork mounts have lateral counterweights. If that is not the case, there is often enough room to drill a hole in the arms, allowing one to screw in a removable counterweight. A drilled square allows the adjustment of the location of the counterweight. This adjustment has to be performed when all the imaging equipment is already in place or lunar tracking will be disturbed, with some risk for the mechanism and the motors. The case of asymmetric forks with just one arm is more awkward. A possible solution is to drill a hole for a counterweight screw in the side of the arm, and, at the opposite side of the mount, screw a shaft for the other counterweight. In all cases, the arm of a fork is not bare: it may harbour wires, a motor, and gears for the declination axis. The base comprises the power supply, at least one motor, wires, and the electronics. Of course, any user modification invalidates the guarantee. That is why we have to take care of the balancing prior to buying a fork mount. Fortunately, if the optical tube has no thread and the fork cannot be drilled, there is a good way to perfectly balance the telescope: purchasing two rings for the optical tube, with both having a Kodak thread for piggy-back installation. This provides two reliable supports for a shaft with a sliding counterweight, with no modification at all of the telescope or the mount.

Azimuthal (also called "altazimuthal") forks are always balanced, with no awkward configuration. Nonetheless, the imaging equipment may be heavy and could destabilize the optical tube. The best solution is, once again, adopting a longitudinal shaft for a shifting counterweight. Lateral balance is not a concern at all because the mount is always straight.

German-type mounts are very easy to balance (Figure 4.14). Either the optical tube can slide within the two rings, or the dovetail can slide along the head of the mount. Optical tubes are seldom directly screwed into the mount, but this may occur with debut, 90–125-mm/4–5-in telescopes. A simple, heavy dew shield made of PVC can counterbalance the mass of the imaging equipment. In other cases, the two axes are balanced by longitudinal sliding of the optical tube (in the mounting rings or with the help of the dovetail) and lateral sliding of the counterweight along its shaft.

4.10   Reducing vibrations of a Newtonian telescope

A Newtonian telescope has a large and long tube and a huge wind surface area (except for truss tubes): this implies more inertia than for other optical setups with the same diameter. In other words, any subtle vibration is magnified by the mass and size of the telescope. A Newtonian telescope on its German-type mount is like a heavy and cumbersome mass rotating on two axes. Basically, the inertia of a

Figure 4.14 The German-type equatorial mount allows a proper balancing in the right-ascension axis by sliding the counterweight (horizontal arrow), and in the declination axis by either translating the tube (vertical arrow) within its mount rings or sliding the dovetail along the head of the mount. Balancing must be perfomed when all the imaging equipment is in place.

rotating mass simply depends on the mass, the radius, and the rotation speed. It is expressed by

$$\text{Inertia force of the rotating telescope} = \text{Mass} \times (2 \times \pi/\text{Duration of one turn})^2 \times \text{Radius}$$

In practice, any gust or little shock causes the optical tube to start rotating. This rotation then becomes a vibration because the tube, which is somewhat elastic despite being fixed on its locked axis, is immediately drawn back, then forward a little less, and so on, and this lasts for several seconds: this is a damped oscillation (Figure 4.15).

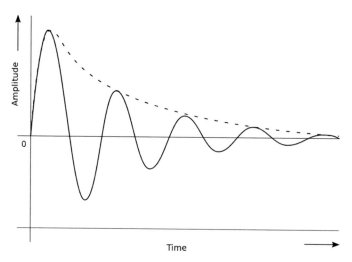

Figure 4.15 This curve shows the amplitude of a damped vibration. All telescopes, especially Newtonians, are prone to vibrate, especially if they have a long tube and a slender, German-type mount. The duration of the phenomenon is generally 3–5 s after each gust or shock. Massive mounts stabilize large optical tubes. Image by Guy Inchbald (Wikipedia).

We can now play with the parameters. If the wind speed is halved, the force of the vibration is four times smaller, but we cannot control the wind. If the counterweights are brought back from 50 cm to 25 cm away from the center of mass, the force is halved, and then the motion blurring is reduced by a factor of two, not to mention the reduction of the duration of vibration damping. We can conclude that the counterweights should be as close to the center of mass as possible. Of course, this implies modifying the mass distribution of the tube, too. One solution is to adopt a carbon tube. Such a tube is light (and expensive). Some amateur astronomers have swapped their steel tube for a carbon tube, but the mass gain (about 1.5 kg or 10% for a 254/1200) was not really appreciable.

Fortunately, a Newtonian telescope can rotate inside the mounting rings. If we orientate the focuser with the imaging train toward the counterweight shaft, the mass of the imaging equipment will be brought closer to the center of mass. This allows the placement of the counterweights closer to the center of mass as well. Thus the inertia of the whole telescope is reduced, minimizing the effects of gusts and vibrations when adjusting the focuser. The drawback is a more uncomfortable body posture for observing.

4.11 Managing turbulence and cooling

Turbulence is the worst enemy of high-resolution imaging. Different layers of air have different, ever-evolving temperatures and densities. Thus the layers of air

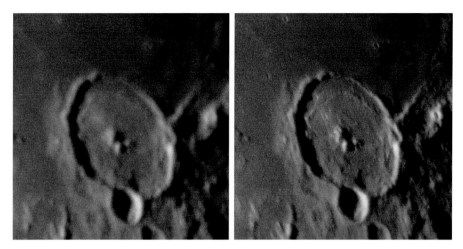

Figure 4.16 Effect of turbulence on two raw frames from the same movie. High-throughput cameras, that is, cameras with a high number of frames per second, are favored because, since the turbulence often acts discontinuously, they are able to acquire numerous frames during unpredictable, calm intervals. The worst frames are abandoned during stacking. Note that some parts of the image are sharp, whereas other parts are blurred (40° W, 17° S).

act as a set of perpetually deforming lenses (Figure 4.16). That is why the magnification is always limited, and why large-aperture telescopes rarely reach their theoretical resolving power. In practice, amateur telescopes do not often show sharper images if their aperture exceeds 300–400 mm (nonetheless, the gain in contrast and light gathering is undoubtable). The best option is to find a high-altitude observation site, as used for professional observatories, to minimize the effect of the atmosphere, because the air density decreases geometrically. Amateur astronomers have to find other strategies to further their struggle against turbulence.

In the real world, there are three kinds of turbulence, and two of them are really manageable:

- air turbulence at different altitudes;
- local turbulence due to the immediate environment;
- internal turbulence of the telescope.

Medium- and high-altitude turbulence is due to perpetually evolving air layers of different speeds, densities, and temperatures. This isn't something we can act on. But we can observe clouds and wind, knowing that misty weather can indicate a steady atmosphere, and we can survey the weather and pressure fronts with the help of websites and weather satellites (Appendix 2). Twinkling of stars is an excellent means to evaluate the stability of the atmosphere. If stars above 45°

Figure 4.17 Each time the light travels through the optical tube, it is affected by internal turbulence and the flaws of each optical part, and suffers a slight loss in light reflection (the typical reflectivity of a primary or secondary mirror is 92%) or transmission (typically 98%–99% for excellent optical parts). Turbulence affects the light path only once in a refractor, twice in a Newtonian reflector, and three times in a Cassegrain, Ritchey–Chrétien, or catadioptric telescope.

clearly twinkle, the atmosphere is unstable and a strong magnification will just show blurred, dancing images.

Internal turbulence can be lowered by cooling. When the air inside the tube is at the same temperature as the external temperature, no convection occurs. This is particularly important with catadioptric and Cassegrain-family telescopes because the light paths cross the same air several times (Figure 4.17).

A speeding up of cooling can be achieved by forcing a flow of air with the help of a fan. We have to consider different types of telescopes.

For sealed-tube telescopes, the following statements apply.

- Compelled cooling of sealed telescopes with thin corrector plates, such as Schmidt–Cassegrains, is not a very good idea, because introducing cool air with a fan, even through a fabric interposed as a dust filter, drags particles into the tube, onto the two mirrors, onto the internal face of the plate corrector, and onto the prime focus lens of the optional field corrector. A serious pull-and-drag system with multiple fans and dust filters is required.
- Maksutov and Rumak telescopes are very hard to cool down because they suffer from the problems associated with all sealed-tube telescopes in addition to the big problem of cooling down a very thick piece of glass: the Maksutov meniscus. No definitively satisfying solution has been reported to date, and precisely this is the main problem regarding these telescopes, notwithstanding their close-to-perfect adequacy for lunar imaging. The main concern is that the cooling time of the meniscus can greatly exceed the time interval during which the air temperature stays steady. If the temperature changes during the night, the meniscus may never be in thermal equilibrium. This is one reason why they will not replace apochromatic refractors for lunar imaging even if the diameter is comparable.
- Refractors are less sensitive to temperature, but they have to cool down too. Their cooling is achieved by exploiting the thermal conductivity of the tube, which is generally made of aluminum, or sometimes steel. Since carbon tubes are excellent insulators, they cool down more slowly.

Problems have been solved with fans, filters, and slots (e.g. Celestron's Edge HD) or filters and slots (e.g. Officina Stellare's Hiper APO). Some DIY enthusiasts have made similar improvements to their sealed, catadioptric telescopes.

For open-tube telescopes the situation is as follows.

- Newtonian, Dall–Kirkham, Cassegrain, Cassegrain–Mangin, and other telescopes experience convection currents in the tube. The main heat source is the main mirror, and, if the latter is reachable, it can be cooled with the help of a simple fan to drag warm air away from the back of the mirror (if the mirror is close to humid grass or dusty soil) or to blow cold air to the back of the mirror (a more efficient solution). A simple 12-V fan can be cannibalized from a desktop.

Large Newtonian telescopes can receive a fan above the center of the aluminized side of the main mirror (this part of the mirror is masked by the secondary anyway) or a fan can be attached to a side of the tube, at right angles and just above the aluminized side of the mirror. Note that a stable boundary layer forms on the surface of the mirror, and destroying it is not a good idea. An open truss tube favors natural cooling and allows convection currents to escape.

4.12 Video-assisted collimation

Collimation, the alignment of the optics, ensures that the telescope reaches the top level of its resolving power – its capability to concentrate light. Various methods and tools exist.

- For all telescopes: observing the diffraction pattern of a star at very high magnification.
- For reflectors: observing the position of a small ring on the main mirror through a Cheshire eyepiece, or observing the position, on the main mirror, of a spot emitted by a collimation laser through its integrated Cheshire eyepiece (Figure 4.18).

In our opinion, the latter methods mainly guarantee the alignment of the eyepiece holder, and any minuscule shift of the focal train (the laser, its integrated Cheshire eyepiece, and the drawtube) alters the position of the spot on the main mirror. Simply tightening the locking screw of the drawtube alters the position of the spot by several millimeters. This is a handy way to prepare the collimation in daylight, but the results are not accurate enough for high-resolution imaging. Anyway, the proper alignment must be done with the definitive imaging train: the filter or filter wheel (if any), the Barlow lens or the eyepiece in the tele-extender, the possible spacers, and then the camera. After all that, as we plan to image the Moon, we need to put the imaging equipment in place. The method has proved to be both efficient and extremely comfortable, especially with long optical

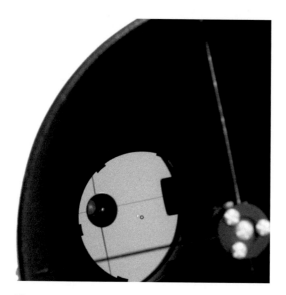

Figure 4.18 Collimating with a Cheshire collimation laser during daylight saves time. An accurate collimation, resulting in sharp images, should be performed with the help of a real star.

tubes, the use of which otherwise implies constantly moving to adjust the colli-mating screws, then going back to the focuser, and so on.

Before doing this in practice, we have to know about some of the differences among the main optical configurations. Apart from the diameter, they precisely guide the choices to be made before purchasing an instrument for lunar imaging.

- A refractor is generally collimated and stable. Possibly adjustments could be made by a professional optician; however, not only does this require specific equipment, but also the lenses are often held in place by lateral screws: the operation is feasible but impractical. No amateur can adjust a triplet or an oil-spaced objective. Some objectives (e.g. Vixen's authentic Fraunhofer achro-mats – with the front, crown lens having asymmetrical curvatures) are aligned with three thin wedges 120° apart. Aligning the whole objective cell is possible, but aligning just one lens is somewhat difficult.
- A Schmidt–Cassegrain or Rumak telescope may need to have its secondary adjusted, generally with a hex key for Allen screws (or handy Bob's knobs, which require no tool). Some models allow adjustment of the primary too.
- Pure Maksutov–Cassegrains generally need no collimation at all because the secondary is a fixed, aluminized part of the meniscus (although some models allow the adjustment of the primary).
- Newtonian/Dobsonian: the secondary can be aligned during daytime with a hex key. The primary needs no tool.

- Some Ritchey-Chrétiens are provided with collimation scopes. Once again, owners tend to ignore them because star testing is more accurate.

In this section, only the popular Schmidt–Cassegrain and Newtonian telescopes are mentioned. Video-assisted collimation can be generalized to other types.

4.12.1 Preparing the collimation

Before operating, here is some advice.

- Adding a red filter diminishes the turbulence.
- When the imaging train is in place, it is wise to aim at a star not too far from the Moon to prevent mechanical and optical bending. We have to let the telescope cool for about an hour while it roughly tracks the star.
- The star must always be re-centered after any slight adjustment. The stellar diffraction image has to be symmetrical only when it is at the very center of the optical axis.
- When the turbulence is too strong, collimation is impossible and useless.
- It is wise to verify the collimation several times per imaging session, especially if the telescope was flipped when its equatorial mount passed the meridian.
- Telescopes that focus by translating the main mirror (Schmidt–Cassegrain and Maksutov–Cassegrain) and common focusers (refractors and Newtonians) may cause image shifting when the focuser knob is turned in one direction and then in the reverse direction. During the collimation and for future image acquisitions, focusing should always be performed from the same point, turning the focuser knob in the same direction. This ensures a better repeatability of mechanical slack, hence the alignment of the optics and the imaging train remains reliable in a given situation.

4.12.2 Collimating a Newtonian telescope

The secondary mirror is best aligned in daylight. This step is easy, and it requires just a screwdriver or a hex key. Some amateurs replace the screws by handier, longer models with a knob.

Once the secondary has been set correctly (Figures 4.19 and 4.20), we have to adjust the primary by night. After the imaging train has been put in place and the telescope has cooled, we can orientate the laptop toward the primary where we stand. The locking screws have to be loosened (Figure 4.21).

Now we can observe the slightly defocused star on the screen, and then, as the collimation gets better, we can focus it. The star must be perfectly re-centered after each eighth of a turn of a push–pull screw (Figure 4.22). The most convenient way is to locate the screw related to the direction of the greatest thickness of the asymmetrical pattern of the star. The rotation of the camera and the possible

149

Figure 4.19 A view through the empty drawtube of a Newtonian during daylight. At the center stands the dark disk formed by the reflection of the secondary (with the four vanes of the spider) on the primary. The large, bright ring is the primary. If the secondary is aligned, that is, with the correct distance to the primary, a correct orientation toward the drawtube and a correct tilt, we can see a symmetrical image and the entire primary with its clips if we stand close enough to the secondary.

Figure 4.20 The secondary of the Newton is to be aligned in distance from the primary by the central screw labeled "D." The secondary can be rotated. These two settings are to be performed while observing the secondary through the empty drawtube. The labels "O" indicate the screws for tilt.

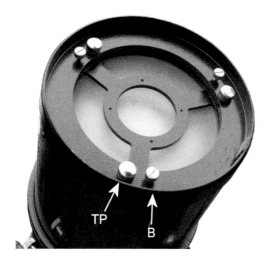

Figure 4.21 The primary has three push–pull screws labeled "TP." The "B" screws are for locking: they have to be loosened prior to collimating the primary, then tightened once the collimation has been performed.

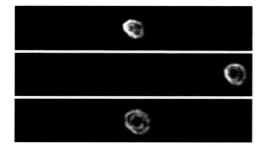

Figure 4.22 Top: the primary is decollimated. The camera is here aligned so that the thinner side of the image corresponds to the leftmost screw, when we face the primary. Center: The leftmost push–pull screw is gently tightened. The star shifts to the right. Bottom: the leftmost push–pull screw if properly set. The star is re-centered with the help of the hand controller to ensure that the collimation is correct at the very center of the optical axis. With experience, video-assisted collimation takes only two minutes.

mirroring of the image with the acquisition software provide large degrees of freedom to get a comfortable correspondence between the image shifting and the position of the first screw, but the camera has to remain in place for the easier, following adjustments, in order to keep the imaging train aligned.

Finally, the locking screws can be tightened. This affects the position of the star, but, once the three locking screws have been tightened, the star can be assumed to be centered again and still symmetrical.

151

Figure 4.23 The three Allen screws tilt the secondary. We have to prevent the secondary from falling by ensuring that at least one of the screws remains moderately tightened, so that the secondary remains against the corrector plate, although it must not compress the plate. The secondary also has some freedom in centering because the diameter of its support is narrower than the hole in the plate: it is precisely adjusted by the manufacturer, along with the rotation of the corrector plate. In addition, the possible rotation of the secondary may affect characteristics such as astigmatism. Normally, only the secondary has to be collimated.

4.12.3 Collimating a Schmidt–Cassegrain telescope

Only the secondary has to be aligned. Once the imaging device has been put in place and the telescope has cooled, the laptop has to be orientated toward the secondary. We adjust the screw by an eighth of a turn or less: generally, a gentle and slight adjustment is sufficient (Figure 4.23). We have to be careful because the hex key may fall on the corrector plate. Large, manually operable knobs are more practical.

The star must be re-centered after each adjustment. This is why facing the secondary while we control the image on the screen is so comfortable. Unlike with Newtonian telescopes, adjusting the secondary of Cassegrain-family tele-scopes requires standing in front of the tube, causing thermal disturbances. If our hand is warm, some tens of seconds are required for the air to cool down before we can carry on.

4.13 Mechanical and optical bending

If the tripod is installed on damp ground, it tends to subside, causing a misalign-ment relative to the celestial pole and a field rotation. If the camera has a very large number of photosites and a slow throughput, this leads to an improper super-imposition of stacked frames, and hence more noise at the edges. Wind also affects tracking, and frame registration during stacking is affected, with comparable results (Figure 4.24). Steel and aluminum tubes are slightly elastic, and they may

Figure 4.24 An unprocessed, stacked image taken under windy conditions. The black areas correspond to missing data and incorrect superimpositions because of gusts. The edges contain a reduced number of frames, leading to a poor signal/noise ratio and lower resulting resolution, while the center may be accurate (20° W, 22° S).

expand or contract with temperature variations. Some elasticity also comes from the tube rings, especially if they are placed on a thin dovetail.

Simple remedies are replacing the dovetail by a larger one (Losmandy standard) and adding stiff tube rings. However, this increases the load and the lever effect. Another solution was exploited for ancient, long refractors placed on German-type mounts: Hargreaves' strut (it stiffens the assembly along a single plane, whereas a Serrurier truss tube stiffens the assembly in three dimensions). A shaft connects the front of the tube to the counterweight shaft to shorten vibrational damping. This method was developed for long refractors, stiffening only the pitch axis, and has to be adapted for larger reflector tubes, which also have a roll axis. Two struts are to be attached at the same point on the counterweight shaft, and attached to each side of the front of the optical tube. Unless the telescope remains in an observatory, the struts must be removable, secured with dowel pins. Because the counterweight shaft rotates with the tube, the attachment is simple and can rely on the thread at its end (which is normally used by the security bolt to retain the counterweights), a clamp, or a gimball.

Large (300-mm/12-in) thinned mirrors tend to bend when the tube crosses the meridian, like heavy imaging trains in awkward configurations. Abruptly aiming a thin 300-mm mirror from East to West blurs the image: it takes 15 minutes to recover its shape. Most telescopes, especially Schmidt–Cassegrain ones, are

Figure 4.25 A slight dew deposit on optical surfaces leads to diffusion of incoming light. This decreases contrast and blurs the image. A more substantial deposit makes focusing impossible. This image was moderately altered by dew on the secondary mirror of a Newtonian telescope; comparable effects occurred with the corrector plate of a Schmidt–Cassegrain telescope. In addition, dew can form on the filter, on the protection window of the camera, or on the Barlow lens. The center of the image has been corrected by drastically lowering levels in the darkest areas; nonetheless the smallest details were blurred (38° E, 56° S).

affected by shifting of the primary while focusing or changing the orientation, as are mirror cells in other kinds of reflector.

These mechanical and optical bendings – not to speak of temperature variations – lead to the need for regular adjustments both of the collimation and of the focusing.[3] Imaging sessions may require ten focusings per night and two or three checks of the collimation.

4.14 Avoiding and eliminating dew

Mist forms when the glass of the optics reaches the temperature at which a gas (air and water vapor) condenses to form droplets – this is the dew point. Contrast drops and light transmission or reflection dims (Figure 4.25). This affects lenses and corrector plates (refractor objectives, Schmidt–Cassegrain, Maksutov . . .), secondary mirrors (Newtonian . . .), and Mangin correctors (Klevtsov . . .). There are preventive and curative remedies.

---

[3] Some particular telescopes use invar (e.g. discontinued Lichtenknecker flat-field cameras) or complementary-expansion-rate materials to avoid expansion/contraction. Carbon is another solution.

Figure 4.26 An unregulated heating resistor for 200-mm (8-in) Schmidt–Cassegrain telescopes; it can be used with smaller-aperture telescopes, refractors, and telelenses. It is to be placed just behind the objective cell or the corrector plate.

- Actively keeping the optics warmer than the dew point with the help of a heating resistor.
  - A heater strap is wrapped at the closest possible location around the photo-lens, objective, or corrector plate, behind it on the tube (Figure 4.26).
  - A flexible or rigid (PCB) heater attached behind the secondary mirror of a Newtonian or Cassegrain telescope. Heaters are provided as elliptical surfaces, plain or annular, or round and annular for a Cassegrain, a Ritchey–Chrétien secondary, or a Mangin secondary/corrector. They require a power supply (12 V, 1 A or more, simple or regulated). The efficiency is excellent and no additional turbulence is noticeable.
  - Disposable or reusable chemical heating pads are wrapped around the photolens or the objective.
- Passively keeping the optics sheltered from humidity with the help of a long dew shield. It may be the light shield provided with the instrument, but this part is generally too short. A convenient length is about twice the diameter, in order for it to remain efficient even with high elevations of the Moon. It may consist of cardboard, foam camping mattress, ABS . . .
- Using simultaneously active and passive solutions for sealed telescopes with a heated dew shield.

A problem exists with carbon tubes. Carbon is an excellent insulator, and it cancels out the action of a heating resistor. Furthermore, a massive apochromatic triplet or a large meniscus is hard to heat. In this case, a long, passive dew shield is a convenient solution.

5

Wide-field lunar imaging

5.1 The Moon in a landscape

Recommended imaging device	Recommended optics	Shooting difficulty	Processing difficulty

As the Moon is generally the brightest element of the scene, the exposure has to be adjusted on it rather than on the lansdcape (Figure 5.1). Street lamps may cause a negligible overexposure because the main subject is the Moon, attracting the eye in the picture. The surroundings may be severely underexposed. In daylight, the entire scene can be correctly exposed, at the price of a strong loss in contrast. If the photography is black and white, a red filter (Kodak W92/W23) dims the daylight and increases the contrast, but this needs a longer exposure.

As usual, automatic exposure and autofocus have to be set in "spot" or "center" measurement while we precisely center the Moon. Some cameras allow us to keep the measure while we press the shutter-release button at mid-course, hence we can re-frame the scene. The flash must be neutralized. In full automatic mode with a short-focal-length photolens, the camera may refuse to shoot because it cannot manage a correct exposure and focusing. The solution is to re-frame to include a large part of the landscape. Advanced techniques like HDR/WDR and tone mapping may help to correctly shoot a small, very bright spot on a dark background.

Figure 5.1 The Moon taken with DSLRs under very different conditions (the images are cropped). Left: the foreground is underexposed while the Moon is overexposed. Center: a composite (4 ms and 1 s) improves the result, but the foreground still resembles a shadow theater. Right: the landscape is somewhat underexposed, but remains readable, while the brightness of the almost full Moon, at low altitude, is attenuated by the misty atmosphere (this is a single exposure, but the areas of low light levels were brightened with an image editor).

5.2 Lunar halo

Recommended imaging device	Recommended optics	Shooting difficulty	Processing difficulty
		☺	☹

When clouds contain small ice crystals, they decompose light like a prism, forming lunar halos and paraselenes ("moon dogs"). These phenomena, which are uncommon though not rare, are visible with the unaided eye and constitute

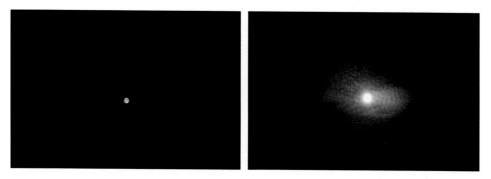

Figure 5.2 Left: an exposure of duration 1/100 s, with the iris set to F/16, where the sensitivity is 400 ISO. The Moon is correctly exposed: maria are detectable. Right: a 1.3-s exposure, with the iris set to F/5.6, at the same sensitivity. The paraselene is visible, but the Moon is overexposed.

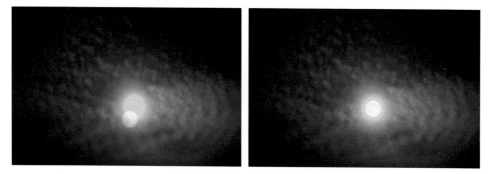

Figure 5.3 Left: the longest exposure is the background layer. The shortest one has an opacity of 50% to align it with respect to the background. Right: a diffuse selection (Section 8.2) keeps the part of the upper layer containing the correctly exposed Moon. Then the contrast is corrected. It is impossible to abruptly cut the correctly exposed Moon because of the presence of the diffusion of the lunar disk in the background layer: this would result in an ungraceful appearance.

beautiful photographic subjects. The very strong contrast is barely manageable even if the camera has HDR/WDR capabilities. Two or three different exposures are sometimes necessary, with substantial bracketing, that is, different settings of iris and/or exposure. The shortest exposure is for a correctly exposed Moon; the longest are for intermediate and low levels of the halo.

Now we can superimpose the exposures with image-editing software (Appendix 2). Figure 5.2 shows two exposures; they were sufficient to record all of the interesting features of the scene within a limited time while clouds quickly passed. The longest exposure forms the background image while the shortest one is transformed into a semi-transparent layer (Figure 5.3), resulting in a well-balanced, composite image (Figure 5.4).

Figure 5.4 After the shortest exposure has been cut with diffuse selection and its contrast corrected, the Moon seems natural again. Nonetheless, the saturation of the background layer has been substantially increased to emphasize the prismatic colors of the paraselene. HDR software resulted in comparable – or worse – results with this image. The brightness of the lunar disk prevented one from preserving the visibility of the clouds passing in front of it. The image was cropped.

Figure 5.5 This lunar halo required a 4-s exposure. The camera was simply attached on a tripod. The image file type was set to raw mode in order to keep all the dynamics. Despite the 18-mm focal length, the field was somewhat too narrow.

Sometimes, a full ring encircles the Moon: the 22°-large lunar halo.[1] The phenomenon lasts some tens of seconds to several minutes. The procedure is

---

[1] There is a slight, differential deviation of 0.8° between colors; the internal edge of the circle is reddish while the outer edge is bluish.

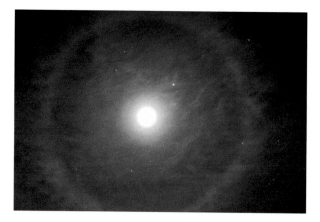

Figure 5.6 The previous image of the halo was processed with the inflexion-point method. A wavelet sharpening improved the readability of subtle gradients. The bright spot near the Moon (upper right) is Jupiter. Other spots are stars: Procyon at the bottom, Castor and Pollux top left, and Betelgeuse with Orion on the right. This 4-s exposure recorded the motion of clouds as brush strokes, while Jupiter and the stars were slightly elongated by the diurnal motion.

the same. Once the flash has been neutralized, a wide range of exposures must be taken, for instance from a tenth of a second to four seconds, iris wide open.

We will now work with a single exposure and the curves tool of a common image editor. In the image, the maria are overexposed and diffusion enlarges the apparent size of the lunar disk. The goal is to brighten the halo while keeping the overexposed Moon at reasonable levels. The best method to reach an acceptable balance is the inflexion-point method (Section 7.6): find one knee or more in the levels curve to abruptly differentiate the brightness of the elements in the image (Figure 5.6). One knee is found between the background and the mist. A second one is found under the brightest levels of the lunar disk and above the levels of the ring. High levels are dimmed while low levels are increased. Intermediate levels, between the two knees, are somewhat leveled up. The contrast is preserved because the difference between the background sky, the mist, and the Moon remains strong.

Moon halo, aurora, and inuksuk

Gilles Boutin (Québec, Canada), has been an aurora hunter since 2002, and a northern voyager since 2006, having made nineteen trips to the Nunavik and the Nunavut (Figure 5.7). He is a photographer, an author, and a speaker about auroras. Figure 5.8 shows a rare photograph (single exposure) featuring simultaneously the invading full Moon, a halo, and an aurora.

Figure 5.7 Gilles at work.

Figure 5.8 "Now I stand close to a huge inuksuk. Violent winds constantly sweep the ground, where the remaining snow is hardened. The temperature sometimes reaches –35 °C or –45 °C; fortunately we are appropriately dressed and we limit the exposure of our skin to the air. Above Hudson Strait, the full Moon draws a huge halo within green auroras."

"At thousands of miles from home, the environment is really wild, dangerous, and mysterious; this is the land of coldness and of the Inuit people. One
is welcome here, and landscapes are fantastic: valleys outlined by great
glaciers, sun pillars, solar or lunar halos … It is close to the magnetic
North pole, the entry point of solar particles triggering auroras. I start from
the village of Ivujivik, in the northern Nunavik. In winter, when ice is
formed, polar bears cross Hudson Bay and Hudson Strait to hunt seals. If
the bears attempt to penetrate the village, the school is warned. We travel in
convoys of snowmobiles and our noise keeps the bears at a distance.
Sometimes, an armed Inuk escorts us. Otherwise, we are warned by the
dogs."

Books by Gilles (in French):

Les aurores boréales Québec–Nunavik (GID), 2010
À la découverte du Nunavik (GID), 2015
www.banditdenuit.com

Equipment: Canon 5D and 7D DSLR, fast F/D ratio, photolenses with
focal lengths of 16, 20, 24, and 40 mm.

5.3 Earthshine

5.3.1 Earthshine without a telescope

Recommended imaging device	Recommended optics	Shooting difficulty	Processing difficulty

A few days before or after the new Moon, the lunar disk is backlit. We should only
see a small crescent, but the reflection of the solar light from the Earth slightly
lightens the dark part of the Moon: this is the earthshine (Figure 5.9). Technically,
this contrasted but beautiful subject is similar to shooting the Moon in a landscape
or just before or after the totality phase of a lunar eclipse.

Figure 5.9 Venus (top) and earthshine with a DSLR and a 180-mm telelens at F/11. The dark part of the Moon is lit by the reflection of the sunlight from the Earth. The contrast between the crescent and the dark area is always very strong.

It can be imaged in detail with no telescope, but a relatively long focal length is better. The image in Figure 5.10 was obtained with a 180-mm telelens only, but the strong magnification is due to the tiny photosites of a monochrome, planetary camera.

Figure 5.10 Earthshine shot with a 180-mm photolens and a planetary camera on a simple photo tripod. Since the diurnal motion limited the duration of the movie before the Moon drifted out of the field, only 140 images were stacked. The dark side was not excessively drowned by the lit, inevitably overexposed crescent. Features are easily visible on the dark side: maria, large craters such as Copernicus, Tycho and its rays, Aristarchus, Plato . . .

Earthshine

Dani Caxete (Spain) is a graphic artist and a photographer (Figure 5.11). His work is based on a great sense of aesthetics and meticulous preparation (Figure 5.12) with relatively modest and standard equipment.

"My great passion is to shoot astronomical events in the frame of landscapes. These kinds of scenes may present technical complications, but I think it is a nice way to bring astronomy to people with little knowledge of astronomical ephemerides. Another passion is to catch the International Space Station in transit over the Sun or Moon, for which I prepare each shoot with great care and precision. When you get good results, photographs never cease to amaze."

Dani was three times a finalist in the Astronomy Photographer of the Year competition organized by the Royal Observatory of Greenwich; he has also been published three times in NASA's APOD (Astronomy Picture Of the Day), twice in APOD calendars, fifteen times in Chuck Wood's LPOD (Lunar Picture Of the Day), and by the BBC, the *Daily Mail, Discover Magazine*'s Bad Astronomy blog, The World At Night, and *Nikon Magazine*.

Figure 5.11 Dani at work.

Figure 5.12 Earthshine with a DSLR and a 300-mm zoom.

Dani's images:

www.flickr.com/photos/danicaxete
www.twanight.org/newTWAN/guests_gallery_c.asp?Guest=Dani%
20Caxete
http://danikxt.blogspot.fr/
http://danikxt.com.es/fotografia/inicio.html

5.3.2 Earthshine with a telescope

Recommended imaging device   	Recommended optics 	Shooting difficulty 	Processing difficulty 

A single exposure may be slightly enhanced by increasing the lowest levels while compressing the highest levels, but this results in noisy images. If the dark side is correctly exposed, i.e. when the main features are distinguishable, the light side is clipped and diffuses light all around. On the other hand, a correct exposure of the bright side excessively increases the noisy and posterized areas of the dark side. A planetary camera or a DSLR with HDR/WDR capability is best suited to earthshine, but manual HDR is feasible.

A mix of at least two exposures helps to get a more acceptable balance (Figures 5.13 and 5.14). In image-editing software, the longest exposure forms the opaque background. The shortest exposure is superimposed as a semi-transparent layer, correctly aligned, and then the layers are fused. Numerous superimposition modes or effects are provided by the software; the best method is to try all of them. For instance, the "hard light" layer mode (PaintShopPro™ and Photoshop®) performed well in this case. The purpose is to mitigate clipping (long exposure) and underexposure (short exposure).

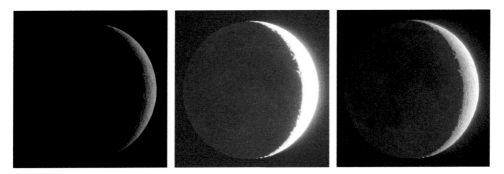

Figure 5.13 Earthshine with a DSLR and a 125-mm (5-in) Schmidt–Cassegrain: 4-ms exposure (left) and 500-ms exposure (center). The frames were manually aligned, and then superimposed as layers to reveal the features both in illuminated areas and in dark areas (right).

Figure 5.14 Earthshine close to the first quarter. This image is a superimposition of one 1/40-s and one 1-s exposure. The lowest levels are increased. The dark side remains very gloomy, but some features are still distinguishable: the main maria and the very weak ray systems of Copernicus and Aristarchus. DSLR and a 150-mm (6-in) achromatic refractor.

5.4 Moon, stars, and planets

5.4.1 Moon, stars, and planets without a telescope

Recommended imaging device	Recommended optics	Shooting difficulty	Processing difficulty

(cont.)

Recommended imaging device	Recommended optics	Shooting difficulty	Processing difficulty

The human eye is said to have a dynamic range of about 100 dB, compared with the 72 dB of recent sensors. The gap is enormous: in practice, an ordinary, still camera needs about 1 s to gather the light from the stars, whereas the unaided eye sees them instantaneously. More important is the fact that the Moon demands short exposures, for instance 1/100 s, while the eye perfectly manages the difference, so that we can see both the Moon and the stars. A camera cannot do this – even with HDR – because the lunar light spreads, resulting in a white glow. The best moment is a lunar eclipse, even in a partial phase, because the dynamic is strongly dimmed.

Fortunately, the Moon may have bright, temporary neighbors such as Mercury, Venus, Mars, Jupiter (Figure 5.15), Saturn, and some bright stars or great star clusters like the Pleiades. A thin lunar crescent might not be overexposed even with exposures of duration 0.25 s or longer when it is low and atmospheric absorption is relatively strong. Other opportunities are provided by conjunctions, when the objects appear in a narrow field of view. Powerful telelenses are required.

A camcorder, too, can shoot such a scene, with a lower resolution (Figure 5.16). The digital zoom is totally useless. An optical zoom of 10× or more is required.

Nonetheless, an ordinary, short-focus photolens brings half the solution. A short focal length is unable to reveal details on the Moon but, due to this short focus, the combination of a wider field and less sensitivity to diurnal motion gives a better chance of being able to image bright stars in the scene, especially when the Moon is a thin crescent dimmed by the atmosphere at low elevation (Figure 5.17).

Details on planets are out of range of such optics. They require a substantial focal length and a motorized mount. A smartphone has a very wide field; hence

Figure 5.15 Jupiter (real size in the green circle, and enlarged by 500%) and the Moon. The image has been vertically cropped. The field of the telelens perfectly fits the scene, but, if the two objects are too distant, we may need to wait for the Moon to move eastward during the night or on the next night. DSLR and an 18–105-mm telelens set to its maximal focal length.

Figure 5.16 Jupiter (green circle) and its 500% magnification with the Moon. The image has been vertically cropped. HD camcorder, with the zoom set to its maximum focal length (61 mm). Despite its undoubtable quality, this consumer camcorder intended to record family memories is not designed for handy, manual focusing and manual exposure setting (the Moon is overexposed). Its optical quality does not meet astronomical requirements.

Figure 5.17 Venus and some bright stars appear in this wide-field image (cropped). DSLR with standard, 18–55-mm lens.

the Moon is minuscule. But a webcam fitted to a short-focus photolens is able to image the Moon with Venus or Jupiter (this requires a 12 × 0.5/M42 adapter for photolenses' threads, such as Pentax, or DIY adapters with a drilled photolens back cover).

5.4.2 Moon, stars, and planets with a telescope

Recommended imaging device	Recommended optics	Shooting difficulty	Processing difficulty

During certain conjunctions, the separation angle is narrow enough to strongly magnify the objects while they stay within the field (Figure 5.18). This requires a motorized mount. A single exposure is sufficient because the contrast is attenuated by the magnification, especially with small-aperture reflectors or achromatic refractors. Nonetheless, Saturn is not very shiny (it resembles an average star for the unaided eye), and it demands a brightness adjustment when the image is processed. Jupiter and Mars are clearly brighter: they are easier targets.

5.5 Lunar eclipses

5.5.1 Umbra and penumbra

Owing to the geometry of the terrestrial shadow cone and the refraction of the red part of the solar light by the terrestrial atmosphere, the Moon may experience three kinds of eclipses.

- When the Moon passes only across the penumbra (the external part of the shadow cone), it is simply dimmed with a subtle gradient.
- When the Moon passes across the umbra, the internal part of the shadow cone, it remains somewhat lit by the reddish solar light refracted by the terrestrial atmosphere rather than totally darkened.
- When the Moon partially crosses the shadow cone, it is partially eclipsed and colored, showing an enormous lightness gradient.

The right exposure cannot be determined: the lit part is overexposed while the eclipsed part is very dark. Lightening the latter after the image has been taken increases noise, especially with low-dynamic-range CMOS sensors. If we try to

Figure 5.18 In this single image of exposure duration 1/100 s, Saturn was visible but markedly darker than the Moon. A diffuse selection (Section 8.2) around Saturn made it possible to increase the levels of the precise area in order to obtain an acceptable, homogeneous contrast of the whole scene. On May 22, 2007, the giant Saturn was 3500 times more distant than the Moon. DSLR with an achromatic refractor, focal length 1200 mm. Violet–blue colors have been dimmed by software to reduce the effect of chromatism.

stack images taken with different exposures, the brightest shows a halo larger than the Moon, thus the gradient on a composite/HDR picture could hardly be smooth.

5.5.2 Brightness variation during a lunar eclipse

The terrrestrial atmosphere contains variable amounts of volcanic ashes, pollutants, clouds, moisture, aerosols, and so on. This affects the transparency and refractive index of the air in complex ways. Consequently, solar light which passes through the terrestrial atmosphere is variously dimmed, refracted, and scattered, in such a way that lunar eclipses show unpredictable, huge variations in brightness. Sometimes, the total lunar eclipse is almost invisible, as if it were a new Moon.

In 1921 André-Louis Danjon rated the brightness of a total lunar eclipse on a scale called "L" (Table 5.1), which is still in use, and now also called the "Danjon scale." Estimations in terms of this scale in magazines help one to roughly forecast the required exposures and optics.

Table 5.1 The Danjon scale rates the luminosity (L) of a total lunar eclipse

L	Description
0	Very dark eclipse. Moon almost invisible, especially at greatest eclipse.
1	Dark eclipse, gray or brownish in color. Details distinguishable only with difficulty.
2	Deep red or rust-colored eclipse. Very dark central shadow, while outer edge of umbra is relatively bright.
3	Brick-red eclipse. Umbral shadow usually has a bright or yellow rim.
4	Very bright copper-red or orange eclipse. Umbral shadow has a bluish, very bright rim.

5.5.3 Lunar eclipse with a webcam and short-focus optics

Recommended imaging device	Recommended optics	Shooting difficulty	Processing difficulty
		😐	😊

A webcam with its short-focus objective is unable to magnify the Moon. We have to unscrew the objective and attach the webcam to more powerful optics. The whole setup has to be placed on a tripod or a mount. The results are not very good, because, when the eclipse enters the phase of totality, the scene becomes very dark. The sensor of a webcam is not intended to operate in low-light conditions. We have to set long exposures, and consequently the noise becomes noticeable (Figure 5.19). In addition, the exposure duration is limited by the software, and very "fast" optics (that is, a low F/D ratio, e.g. 1:2.8) is required. A short telelens, for instance 135 or 200 mm in focal length, is perfect, such as a Pentax-compatible telelens from a film camera, with a spacer and a 12×0.5 adapter (Figure 3.6). The spacer is needed in order to reach infinity focus because the back focus of such optics is 45.46 mm, much more than the short-focus objective of the webcam.

Intermediate phases show a strong contrast, comparable to earthshine. We have to choose between a correct exposure of the still lit part of the Moon, or of its dark part, perhaps making a composite image afterwards.

Even if a webcam is poorly adapted to "long" exposures, it offers a good opportunity to measure the size of the Moon relative to the Earth. As the Sun is 400 times more distant from the Earth than the Moon, we can consider that the shadow cone of the Earth is almost cylindrical. The shape of the shadow cast on the Moon is a disk. Since the disk measures almost four times the size of the Moon, knowing the size of the Earth calculated by Eratosthenes (276–195 BCE), we can derive the size of the Moon.

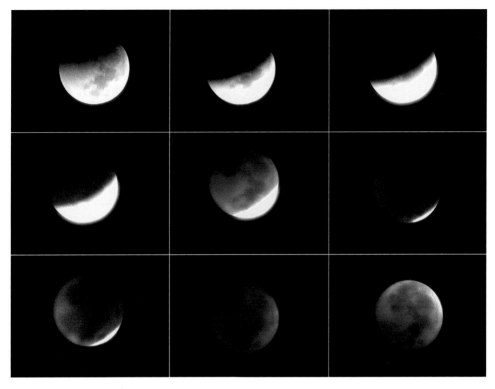

Figure 5.19 Different steps of a lunar eclipse recorded with a webcam at the prime focus of a viewfinder. Note the difficulty in balancing the light and dark areas. The exposure was chosen manually, but an automatic setting gave comparable results. The noise becomes conspicuous during the totality. The objective of the webcam was removed, and then simply attached to a 9 × 50 viewfinder with no eyepiece (these optical parts were simply unscrewed). The equipment was placed on a motorized mount because the exposure duration was as much as 1 s during the totality.

5.5.4 Lunar eclipse with a still camera and short-focus photolens

Recommended imaging device	Recommended optics	Shooting difficulty	Processing difficulty

Figure 5.20 A lunar eclipse during the totality phase; the bright arc to the North indicates that the Moon was tangential to the penumbra. The full Moon is so dimmed that stars are easily recorded at the same time. DSLR, 18–55-mm lens set to 55 mm, 200 ISO, 5-s exposure. The background brightness was increased to improve the clarity.

When the eclipse is total, the solar light is refracted by the Earth's atmosphere and curved (just as a pen viewed when it is partly immersed in a glass of water looks broken). As a result, the Moon is slightly lit, and it shows a magnificent red color, like that of blood or a brick. This red is precisely the color of a sunset. The tint and the lightness vary considerably from one eclipse to another, but the Moon is always strongly dimmed. With the unaided eye, the show is magnificent: the eclipsed Moon resembles the planet Mars, surrounded by numerous stars (Figure 5.20).

This brings a unique opportunity to shoot stars and the reddish full Moon at the same time with an ordinary still camera. The contrast is very low, and all sensors can manage the consequent small difference in brightness between the stars and the eclipsed Moon. But the necessary exposure duration may be rather long, some seconds, and this requires a motorized mount. If the focal length is short, a simple tripod may be sufficient: see the maximal exposures with no motion blurring in Section 5.5.6.

If the focal length is significant, and the exposure duration is long, motion blurring results because of the diurnal motion (in addition to the slight proper motion of the Moon). If we have just a tripod, the iris must be set to the maximal aperture, e.g. 2.8, and the sensitivity has to be set at its maximal acceptable value, e.g. 6400 ISO. The goal is to gather as much of the incoming light as possible in order to get the shortest possible exposure, thus countering motion blurring.

5.5.5 Lunar eclipse with a telescope and a DSLR

Recommended imaging device	Recommended optics	Shooting difficulty	Processing difficulty
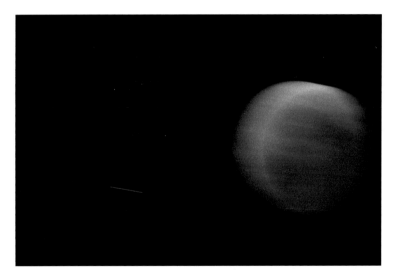			

The telescope is exploited in place of a photolens, and the bare DSLR is attached with a T2 adapter. If the camera is a bridge camera or a compact one, we have to rely on digiscopy. With a significant focal length, celestial mechanics can no longer be neglected, and a motorized mount will be necessary (Figure 5.21).

Figure 5.21 Total eclipse of the Moon with a 30-s exposure and an unmotorized mount: the motion blurring is due to the rotation of the Earth. The focal length is 1250 mm. The oblique line at bottom left is a star trail.

Figure 5.22 A 10-s exposure close to totality (the Moon was not exactly at the center of the shadow cone of the Earth). The motorized mount compensated for the motion: the Moon and the surrounding stars do not suffer from motion blurring. DSLR at prime focus (focal length 1250 mm).

Since the phase of totality requires long exposures – some seconds – the best technique to avoid vibrations that are amplified by the magnification is manual occultation. The DSLR is set in bulb exposure, or "B" exposure: this means that the exposure is unlimited. We place a mask (black cardboard, a lens cover . . .) in front of the objective/telescope, and then, after some seconds, vibrations are damped and we can start the exposure by removing the mask. Then we replace the mask and stop the exposure. A remote controller is generally necessary to start and then stop the exposure, but with some DSLRs one may open the shutter by pressing the shutter button once to start the exposure, and then pressing it a second time to stop the exposure. This manual occultation cannot operate with short exposures. This technique is commonly exploited in deep-sky imaging. A remote controller is extremely useful in order to avoid handling vibrations.

A motorized telescope mount compensating for the diurnal motion is now indispensable (Figure 5.22). Then 10–30-s exposures become possible. We can control the images on the screen of the DSLR to correct the settings (Figure 5.23); fortunately, a lunar eclipse is a slow phenomenon, and we have plenty of time to correct the exposure if necessary. A total lunar eclipse may last four hours.

5.5.6 Maximal exposure, field width, and focal length

Table 2.5 indicates maximal possible exposures for a DSLR with an APS-C sensor of 20 megapixels and no motorized mount.

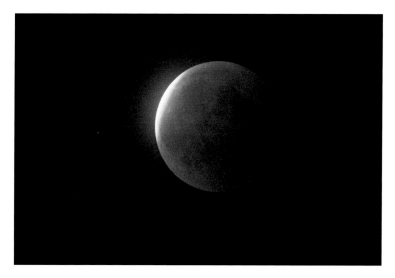

Figure 5.23 Forty minutes later, the Moon emerges from the shadow cone. Tints vary from brick red, at the center, to light gray in direct sunlight. The contrast is now strong, and an HDR capability is required in order to avoid overexposure (Figure 5.24). The single exposure here lasted 5 s. The halo is due to haze. Note the effect of the lunar proper motion: the Moon drifted eastward and the star to the left is now closer. The star to the right on the previous image is now out of the field.

Since the totality phase requires 10–30-s exposures with a light travel mount (e.g. Vixen Polarie, Astro Trac, Vixen Photo Guider, IOptron Sky Tracker, Skywatcher Sky Adventurer, and others) or a light telescope mount. A telelens may be attached by a Kodak screw directly onto the dovetail in place of the telescope, or on one of the attachment rings of the telescope. This requires an approximate polar alignment.

5.5.7 Lunar-eclipse movie

Recommended imaging device	Recommended optics	Shooting difficulty	Processing difficulty
		😐	🙁

(cont.)

Recommended imaging device	Recommended optics	Shooting difficulty	Processing difficulty
(pro movie camera with manual settings)			

5.5.7.1 Brightness variation

The huge variation in brightness of a lunar eclipse is a concern we have to deal with. The first color movie of a lunar eclipse was taken in the late 1970s by Frenchman Pierre Bourge with a 35-mm film camera and a home-made, mechanical automatic iris to compensate for continuous brightness variation (and a home-made, driven equatorial mount). Unfortunately, consumer imaging devices such as camcorders, smartphones, most analog-output video cameras, and compact cameras (when crudely exploited) are internally equipped with a non-disengageable automatic gain control (AGC). An AGC prevents one from correctly adjusting the exposure/iris, leading to overexposed frames.

On the other hand, a camera movie is not necessary. Since a lunar eclipse lasts hours, we have plenty of time to shoot the Moon with a still camera (or a planetary camera), adjust the exposure and iris parameters, shoot again, and so on. One image every 10–300 s is a good interval, according to the file size of the images (or short movies), the number of images (or movies), and the capacity of the storage media. Intervals of 10 s will give a fluid movie; 300-s intervals are used for a typical time-lapse series. We have to adjust the exposure and iris values with the help of the histogram and note them at each step. The movie can be reconstructed later.

5.5.7.2 Preparing the session

A good drill is as follows.

(1) Estimate the overall data volume.
(2) Initialize the imaging device.
(3) Note the parameters during the first half of the eclipse.
(4) Reset them symmetrically during the second half. They cannot be derived prior to the eclipse because the brightness variation is largely unpredictable. Some perturbing factors may arise, and parameters may consequently have to be adjusted.

First, we have to estimate the data volume.

- *For a still image camera,*

Volume $=$ Image volume (megabytes) $\times$ Eclipse duration (minutes)/Interval (minutes)

For instance, with JPG images from an EOS 350 (3000 × 2000 pixels),

$$2.5 \text{ Mbytes (mean volume)} \times 210 \text{ minutes}/1 \text{ minute} = 525 \text{ Mbytes}$$

And with raw images (Canon CR2 format),

$$6.9 \text{ Mbytes (mean volume)} \times 210 \text{ minutes}/1 \text{ minute} = 1.45 \text{ Gbytes}$$

A 14-megapixel DSLR may provide 4-Mbyte JPG images (4500 × 3000 pixels with compression). Owing to its tiny-pixel, CMOS sensor, the number of pixels ("image resolution") may be decreased without noticeable degradation of images. Estimating the mean value for JPG images is important: an image can be complex, with numerous details, or it can have few details but large color gradients, or both. This will change the file size and affect the total amount of data. For example, a "normal" full Moon shows details on a very small part of an otherwise perfectly black image. During a total eclipse, the image taken with a wide-angle lens can comprise the Moon, the landscape in low light, numerous stars, and clouds. All these elements partially affect the size of the compressed image. Consequently, we have to measure the size under various conditions prior to estimating the required total amount of storage. The overall data volume must fit within a single SD card to avoid any need for manipulation at a critical moment of the lunar eclipse.

Video recording is useless because of the reduced format and the limited SD-card capacity (total lunar eclipses last more than three hours!), but a time-lapse recording is perfect.

- *For a camcorder,* the main parameter is the bit rate, that is, the compression. It is either denoted by name, e.g. "XP" for high quality and "EP" for extended duration and lesser quality, or expressed in terms of the bit rate (data flux). The higher the rate, the better the quality and the longer the movie. The menu generally indicates the remaining duration of recording. If not, we have to estimate the remaining duration by recording a long test movie. The menu indicates the remaining capacity as a percentage. The bit rate can finally be adjusted in order to have up to 3 h 20 min left.
- *For a planetary camera,* the device records either images or movies. Uncompressed formats are better: TIFF or FITS images for 10-bit+ cameras, BMP images for 8-bit color cameras, SER movies for 10-bit+ cameras, or uncompressed AVI movies for 8-bit color cameras. Depending on the acquisition software, JPG compressed images can be recorded instead of uncompressed images if the hard disk/SSD has no room left. The JPG compression rate conditions the data volume.

Then, we can set the constant parameters of the device/software as follows.

Table 5.2 Scheduling shoots during a lunar eclipse

Time to maximum of eclipse (minutes)	Exposure and iris (exposures may be multiple in the case of manual HDR or automatic bracketing)
T minus 90	(Note the right values after histogram)
T minus 85	(Note the right values after histogram)
. . .	
T minus 10	(Note the right values after histogram)
T minus 5	(Note the right values after histogram)
T	
T plus 5	Same as T minus 5 min
T plus 10	Same as T minus 10 min
. . .	. . .
T plus 90	Same as T minus 90 min

- DSLR/compact/hybrid still camera sensitivity: a good value is 800–1600 ISO.
- DSLR image format: JPG and/or RAW.
- Camcorder: AVCHD compression ("bit rate"), according to the data-storage capacity of the device, remembering that a camcorder can simultaneously record images *and* movies, particularly if it has separate SD cards.
- DSLR: mirror release.
- DSLR: bracketing.
- DSLR: optional intervalometer.
- Planetary camera: gain and image format, in preference TIFF/FITS image format, or SER/uncompressed AVI movie. During a dark lunar eclipse, it is better to increase the exposure duration rather than the gain, which amplifies noise. The acquisition software may have an intervalometer function, which will need to be properly set.

Finally, we can carry out the procedure described in Table 5.2 to note the variations of shooting parameters during the first half of the eclipse. In this example, the interval is 5 minutes. We can choose a smaller interval next to the maximum (totality or maximum partial eclipse).

Then the same parameters can be recalled in the reverse order during the second half of the eclipse (Figure 5.24). That is, if course, if no typical problems occur, such as clouds, a gradient of air transparency, humidity, variation of the altitude of the Moon, clouds, dew on the lens . . . or a flat battery. The procedure can be exploited with a semi-professional camcorder, a planetary camera, or an "expert" compact/hybrid still camera in manual mode.

For such a duration, if we need a computer, especially for high-frame-rate planetary cameras, a car battery with a DC/AC converter is required. A DSLR needs one or two spare batteries, or a lead–acid battery (Pb) and a DC/DC converter, or a small car battery with a DC/AC converter and a model-specific, AC power supply to enhance autonomy.

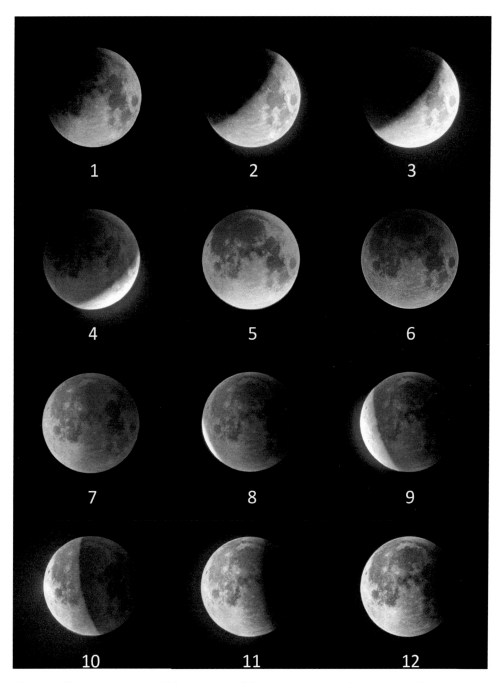

Figure 5.24 The penumbral (1) and totality (6) steps are easy to shoot with single exposures, but highly contrasted, intermediate steps require the assembly of multiple exposures with the help of HDR software (or a standard image editor, with much more work). Images 2–5 and 7–12 resulted from stacking of various exposures of durations from 1/4000 s to 10 s. The end result is quite similar to observations with the unaided eye or binoculars. Images taken on September 28, 2015 with a small refractor and a motorized mount with lunar tracking.

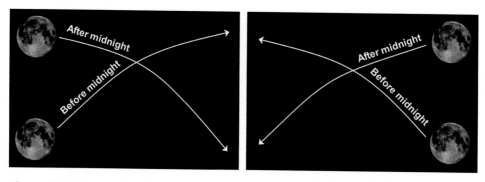

Figure 5.25 Illustration of apparent westward drifting of the Moon in the field of view. Left: northern hemisphere. Right: southern hemisphere. The size of the Moon is exaggerated. From an observation site near the equator, the trajectory is rectilinear.

5.5.7.3 Managing drifting and field of view

Not only do exposure and iris values have to be manually – and continuously – adjusted, but also the diurnal motion has to be taken into account. The additional, orbital motion of the Moon of 0.5° per hour (varying because of its elliptic orbit) may be important for long-focal-length imaging, too. As a lunar eclipse lasts hours, we have to prepare the frame. The diurnal motion leads to an apparent drift of 15° per hour. With its three-hour duration, an eclipse requires either wide-angle imaging or a motorized mount (Figure 5.25).

- If we are in the northern hemisphere, the westward drifting is from left to right when we look to the south (toward the ecliptic).
  - If the Moon is rising (before midnight), the Moon has to start at the bottom left corner of the image.
  - If the Moon is declining (after midnight), the Moon has to start at the top left corner of the image.
- If we are near the equator, the westward drifting is linear, from horizon to zenith (before midnight) or from zenith to horizon (after midnight).
- If we are in the southern hemisphere, the westward drifting is from right to left when we look to the north (toward the ecliptic).
  - If the Moon is rising (before midnight), the Moon has to start at the bottom right corner of the image.
  - If the Moon is declining (after midnight), the Moon has to start at the top right corner of the image.

Note that image inversion must be taken into account, too, depending on the optics, possible accessories such as diagonals, and image inversion on the screen.

During a total eclipse, the diurnal motion is 3 hours × 15° per hour = 45° (we can neglect the proper motion of the Moon). This is quite a large field of view, and a short focal length is required.

Using a motorized mount to keep the Moon in the center of the field allows a high magnification and dramatic images. But this has some significant drawbacks.

- A very sturdy mount, with a perfect polar alignment, is needed. Tracking may be set to lunar speed, but sidereal speed is accurate enough if we keep a safety margin in the field.
- With such a duration, drift is unavoidable. The highest magnification cannot be exploited; so we must, again, allow a margin.
- Manual corrections of drifting, either during acquisition or thereafter during movie editing, will be accelerated and exaggerated. The movie will be jerky.

The author has tried such a solution to film long-duration solar prominences with a solar refractor on a fixed, sturdy Gemini G42 mount and an excellent polar alignment with an adapted Bigourdan method. Nonetheless, small guiding corrections were applied during the three-hour tracking, and they looked like jerks after temporal compression of the movie. Manual re-alignment of single frames was too tedious to be considered a viable solution.

Another possibility is autoguiding, a system exploited for long-exposure, deep-sky imaging. This requires an auxiliary telescope (or an off-axis guider), a guiding camera with included or standalone autoguiding software, and a motorized mount with an autoguider port such ST4. But guiding was developed for stellar imaging, following a star as a precise guiding reference, because a star is a pinpoint. If a bright star is visible despite the brightness of the full Moon (the latter is dimmed when in penumbra or umbra, so the star becomes easier to detect), the mount can accurately follow the star. With a lunar proper motion of 0.5 arcsecond per second, a westward drifting of one lunar diameter per hour occurs, thus a left margin (northern hemisphere) or a right margin (southern hemisphere) of 2 degrees or so has to be saved. And, once again due to the lunar motion, a final 2-degree offset between the Moon and the star must be considered. Of couse, the guiding star has to be to the right of the Moon, or the star will be masked out by the Moon. Autoguiding requires experience, and any problem may halt the operation, such as cable torsion, stopping when passing by the meridian, software crashing … Wise lunar astrophotograhers should put in place a backup imaging system to avoid partial or complete loss of images.

Another solution is inherited from solar photography, which suffers from the same tracking concerns. Keeping the frames steady can be achieved in the aftermath of the observing session with the help of software such as "Dave's Video Stabilizer" by Starry Dave or PIPP by Chris Garry, with the function "Enable Surface Stabilisation" (Appendix 2).

5.5.7.4 Creating the movie

Assembling still images, whether single frames from a still camera or stacked frames from a planetary camera, can be done with the help of movie-editing software: e.g. OpenShot, Virtual Dub, iMovie, Final Cut Pro, Adobe Premiere, Sony Vegas Pro, Microsoft Movie Maker (the latter for short-duration movies), and many others, for various home operating systems; some are cross-platform software.

Fluidity requires a minimum of ten images per second, otherwise the movie can resemble a cheap animation, and the continuously evolving phenomenon could appear jerky. For reconstructing the movie one needs the calibration of timing from the time stamp of each image (file attributes or image EXIF embedded data). Most types of movie-editing software apply a default duration for inserting still images, but this default duration can be modified (via the "preferences" menu). If too few images are available, we may link images with the classic cross-fade. The minimal duration to appreciate a still image is six seconds. The cross-fade may last one or two seconds. Of course, a three-hour movie might be exhausting, and the need for temporal compression is obvious, e.g. a 30× compression to obtain a captivating, six-minute movie from a boring, three-hour lunar eclipse.

Once we have exported the movie, the unavoidable nightmare is transcoding it. The aspect ratio (16:9, 4:3, …) is often unrecognized, or badly interpreted, by conversion freeware: this requires preliminary tests with short excerpts. Standard containers, for instance AVI, WMV, MP4, and common DivX or Xvid codecs are accepted by numerous consumer Blu-ray/DVD/USB flash-memory readers. To avoid having very large movie files, the H264 codec is a very good choice, along with FLV or others, but they can rarely be read on consumer disc players; they need a computer. The free VLC reader (www.videolan.org/vlc/index.html) easily manages various formats and aspect ratios. The last option is to connect the computer to a video-projector with an HDMI or VGA cable.

5.5.8 Lunar-eclipse schedule: 2016–2030

For years, Fred Espenak has maintained two comprehensive websites about eclipses, with precise maps, dates, and all relevant information about eclipses, along with his superb images:

http://eclipsewise.com/eclipse.html
www.mreclipse.com/MrEclipse.html

His webpages also include solar eclipses, but you should never try to image (or observe) a solar eclipse without appropriate solar filters in front of the instrument or photolens. This is a very different technique, with potential danger for your eyes (and your cameras), so it is not discussed in this book devoted to the Moon.

Fred Espenak has also calculated tables and maps for NASA's website (Table 5.3):

http://eclipse.gsfc.nasa.gov/lunar.html

Table 5.3 Lunar eclipses 2016–2030, adapted from eclipse predictions by Fred Espenak, taken from NASA/GSFC

Date	TDT[2] at maximum	Type	Duration (partial in normal type, total in bold)	Visibility
2016 Mar 23	11:48:21	Penumbral	–	Asia, Australia, Pacific, western Americas
2016 Sep 16	18:55:27	Penumbral	–	Europe, Africa, Asia, Australia, West Pacific
2017 Feb 11	00:45:03	Penumbral	–	Americas, Europe, Africa, Asia
2017 Aug 07	18:21:38	Partial	01 h 55 m	Europe, Africa, Asia, Australia
2018 Jan 31	13:31:00	Total	03 h 23 m **01 h 16 m**	Asia, Australia, Pacific, western North America
2018 Jul 27	20:22:54	Total	03 h 55 m **01 h 43 m**	Southern America, Europe, Africa, Asia, Australia
2019 Jan 21	05:13:27	Total	03 h 17 m **01 h 02 m**	Central Pacific, Americas, Europe, Africa
2019 Jul 16	21:31:55	Partial	02 h 58 m	Southern America, Europe, Africa, Asia, Australia
2020 Jan 10	19:11:11	Penumbral	–	Europe, Africa, Asia, Australia
2020 Jun 05	19:26:14	Penumbral	–	Europe, Africa, Asia, Australia
2020 Jul 05	04:31:12	Penumbral	–	Americas, southwestern Europe, Africa
2020 Nov 30	09:44:01	Penumbral	–	Asia, Australia, Pacific, Americas
2021 May 26	11:19:53	Total	03 h 07 m **00 h 15 m**	Eastern Asia, Australia, Pacific, Americas
2021 Nov 19	09:04:06	Partial	03 h 28 m	Americas, northern Europe, eastern Asia, Australia, Pacific
2022 May 16	04:12:42	Total	03 h 27 m **01 h 25 m**	Americas, Europe, Africa
2022 Nov 08	11:00:22	Total	03 h 40 m **01 h 25 m**	Asia, Australia, Pacific, Americas
2023 May 05	17:24:05	Penumbral	–	Africa, Asia, Australia
2023 Oct 28	20:15:18	Partial	01 h 17 m	Eastern Americas, Europe, Africa, Asia, Australia
2024 Mar 25	07:13:59	Penumbral	–	Americas
2024 Sep 18	02:45:25	Partial	01 h 03 m	Americas, Europe, Africa
2025 Mar 14	06:59:56	Total	03 h 38 m **01 h 05 m**	Pacific, Americas, western Europe, western Africa

[2] Terrestrial Dynamic Time (superseded by Terrestrial Time), derived from International Atomic Time. For our purpose, which is a hobby, we can assume that it is roughly equivalent to Coordinated Universal Time (UTC) + 32 seconds.

Table 5.3 (cont.)

Date	TDT[2] at maximum	Type	Duration (partial in normal type, total in bold)	Visibility
2025 Sep 07	18:12:58	Total	03 h 29 m **01 h 22 m**	Europe, Africa, Asia, Australia
2026 Mar 03	11:34:52	Total	03 h 27 m **00 h 58 m**	Eastern Asia, Australia, Pacific, Americas
2026 Aug 28	04:14:04	Partial	03 h 18 m	East Pacific, Americas, Europe, Africa
2027 Feb 20	23:14:06	Penumbral	–	Americas, Europe, Africa, Asia
2027 Jul 18	16:04:09	Penumbral	–	Eastern Africa, Asia, Australia, Pacific
2027 Aug 17	07:14:59	Penumbral	–	Pacific, Americas
2028 Jan 12	04:14:13	Partial	00 h 56 m	Americas, Europe, Africa
2028 Jul 06	18:20:57	Partial	02 h 21 m	Europe, Africa, Asia, Australia
2028 Dec 31	16:53:15	Total	03 h 29 m **01 h 11 m**	Europe, Africa, Asia, Australia, Pacific
2029 Jun 26	03:23:22	Total	03 h 40 m **01 h 42 m**	Americas, Europe, Africa, Middle East
2029 Dec 20	22:43:12	Total	03 h 33 m **00 h 54 m**	Americas, Europe, Africa, Asia
2030 Jun 15	18:34:34	Partial	02 h 24 m	Europe, Africa, Asia, Australia
2030 Dec 09	22:28:51	Penumbral	–	Americas, Europe, Africa, Asia

5.5.9 Occultations

Recommended imaging device	Recommended optics	Shooting difficulty	Processing difficulty
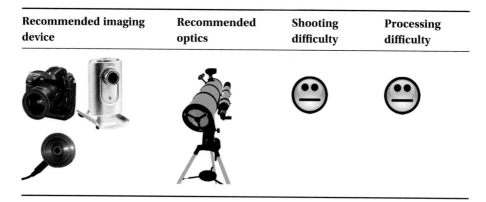			

Owing to its proper motion, the Moon seems to pass in front of distant celestial bodies at a speed of more than one lunar diameter per hour. Sometimes, this body is a bright star, disappearing behind the limb of the Moon, then emerging from the opposite limb: this is an occultation.

When the star passes very close to a lunar pole, it is a grazing occultation. The star blinks as it is masked out by mountains and crater rims, then brightens again when its light passes back into the line of sight, through valleys. This provides an accurate way to measure lunar relief, since no atmosperic turbulence occurs on the Moon, although the image can be blurred by the terrestrial atmosphere. For this reason, occultations are observed, accurately timed, with annotations about the observing location (coordinates and altitude) from differing latitudes, in order to reconstruct the polar relief. Like an eclipse, the area from where a grazing occultation can be observed is long but very narrow: several thousand kilometers long by some tens of kilometers wide.

An occultation is hard to image due to the extreme difference in brightness between the star and the Moon. The human eye easily adjusts to the brightness gradient, but cameras do not. If the star is faint, we must overexpose the Moon, and a bright halo appears, drowning the star. Using an automatic bracketing mode or an HDR-capable camera is useless because of this halo. This is why so few images are published.

The best ways to image an occultation are, at present:

- using a 16-bit, deep-sky camera, whose sensor has a large well capacity, e.g. a Kodak KAF-family sensor (KAF-1603ME, KAF-16803 . . .);
- using an HDR/WDR planetary camera with a large well capacity;
- waiting for the Moon to pass in front of a bright star (this is less expensive but offers fewer opportunities).

Assuming that the apparent motion of the Moon is 0.5 arcsecond per second on top of the diurnal motion, we can consider that we have no more than a single second to take multiple exposures or multiple frames. A planetary camera can take several hundreds of frames to be stacked. A DSLR can take several frames per second or record short movies. But the purpose can sometimes be not to obtain beautiful images, but rather to obtain useful images, or, simply, images. To do so, one frame per second will suffice. Once we have frames, some minimal processing is required in order to compress the contrast. The goal is to brighten the star and, simultaneously, to dim the brightest areas of the Moon. An unsharp mask may be applied later. Figures 5.26 and 5.27 show the emergence of a star of magnitude 3.6 (λ Gemini) on February 1, 2015.

Observing an occultation is more interesting with accurate date and time specifications. A time stamp can be added to each frame after the recording session (e.g. with a Lucam recorder), employing the embedded data of a SER movie, or during acquisition (e.g. with FireCapture). Prior to acquisition, the computer clock must be properly set using a time server.

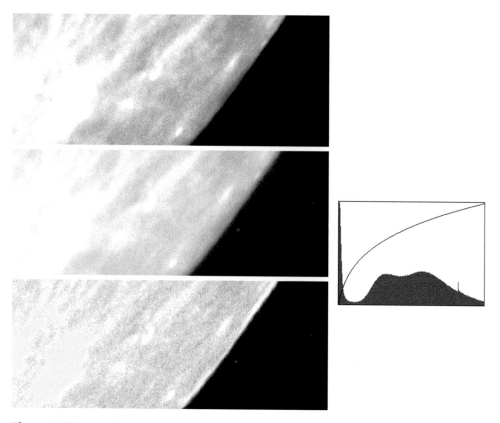

Figure 5.26 Top: close-up of an unprocessed single frame. The Moon is deliberately overexposed. Middle: low levels are increased (see the histogram of the full frame on the right), unveiling the star of magnitude 3.6. Bottom: unsharp mask.

5.5.10 Video time insertion

This technique primarily involves the insertion of a precise time stamp in each frame of a video movie by a device known as a video time inserter (VTI), or video time overlay, connected between the analog-output camera and the recorder. A VTI generally has composite input and output because it is intended for 576-line (CCIR) video cameras. The time stamp comes from GPS satellites. For instance, the classic IOTA VTI (of the International Occultation Timing Association, Appendix 2) is available for $250–300 (depending on options) from various dealers.

As of 2016, most video cameras have digital outputs. In Figure 5.27, the time stamp was directly added to each frame by the image-acquisition software. The computer's real-time clock must be properly set according to a time server. The time stamp must be in UT.

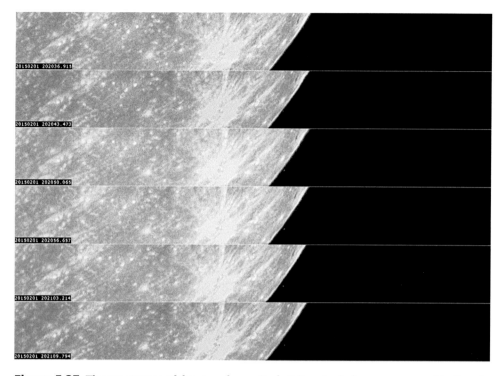

Figure 5.27 The emergence of the star of magnitude 3.6, a single frame per second in time-lapse mode (February 1, 2015). Celestron 5, reducer/flattener F/D = 6.3, cooled ASI120MM, exposure 0.8 ms, gain 60%; the Moon was almost full. Frames are severely cropped; full frames show a quarter of the surface of the Moon. Acquisition with on-the-fly, time-stamp insertion by Torsten Edelmann's FireCapture software.

Numerous still cameras have a GPS receiver to give the precise date, time, and location of each image. Even cameras with no GPS have a real-time clock, but it has to be accurately set prior to the imaging session. The information is kept in the EXIF data. Camcorders also have a real-time clock, and the constant superimposition of the time stamp is possible while recording for certain models. Unlike video time insertion with a GPS receiver, the time from an internal clock is not reliable, and an occultation event that is observed correctly but with no precise time stamp is not very scientifically useful.

6

High-resolution lunar imaging

Recommended imaging device	Recommended optics	Shooting difficulty	Processing difficulty

6.1 Sampling: finding the right magnification

Sampling is like coding music onto a compact disc: good results require both temporal accuracy (to preserve the notes and the instruments' tones)[1] and dynamic range (to preserve the variations in sound pressure, i.e. the volume). The equivalent parameters for images are the sampling rate and the dynamic. The latter is conditioned by the optics and the analog-to-digital converter, possibly with the addition of in-built HDR and WDR capabilities: it can be managed only by increasing the resulting F/D ratio to decrease the dynamic. The first parameter is entirely free: it depends on the size of the photosites and on the magnification. Finding the right magnification involves fitting the size of the details of the image to the size of the photosites.

- If the resulting focal length is too short, or, in other words, the magnification is too small, several details of the image are cast on the same photosite. This results in a lack of image resolution, because the camera is unable to reproduce the full accuracy of the optical image (Figure 6.1, left). This is undersampling.
- If the resulting focal length is too long, each detail is cast on several adjacent photosites (Figure 6.1, right). Not only does this result in a darker image, because the quantity of light is unproductively spread over too large a surface,

---

[1] This refers to the Nyquist–Shannon sampling theorem.

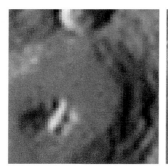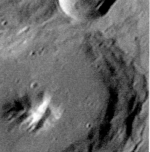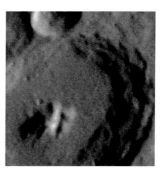

Figure 6.1 Left: undersampling, showing a good contrast but a poor resolution with pixelization. The image has been magnified with a linear zoom (no interpolation) for the comparison. Center: the size of the details of the image is commensurate with the size of the photosites. The resulting resolution matches the optical image's accuracy. Right: oversampling. Here the resulting resolution is correct but the contrast is reduced, no additional detail is visible, the darker image led to increasing the exposure, and the image is more altered by the turbulence. In addition, some photosites were wasted, reducing the field. The image was downsized with linear interpolation for the comparison (32° E, 29° S).

but also the resulting resolution is lower than the accuracy of the optical image. This is oversampling.

- Nonetheless, these two extreme cases have their uses: undersampling provides brighter (less dispersion) and more contrasted (less magnification) images, suited to small instruments with no lunar tracking or with absorbent filters, whereas oversampling may reduce a high contrast that the sensor could not manage.

Given the possible resolving power of a telescope, the right sampling may easily be estimated as half the observational resolving power because there is no reason for each detail of the optical image to be cast exactly at the center of each photosite. Some astrophotographers use one-third the observational resolving power.

Now we have to consider monochrome and color sensors. A monochrome sensor corresponds exactly to the ideal case. A color sensor with a classic Bayer matrix exploits four photosites per resulting pixel, but, thanks to the matrix calculus, it loses only roughly half the resolution relative to a monochrome sensor. Unfortunately, the loss of 20%–30% in lightness because of the matrix prevents the use of more than twice the sampling rate relative to a monochrome sensor, to keep exposure durations reasonable. These considerations refer to the color and monochrome versions of the same sensor.

We can take, as a representative example, a telescope with an aperture of D = 150 mm:

$$\text{Theoretical resolving power} = 120/D \text{ (rough formula)}$$
$$= 120/150 \text{ mm} = 0.8 \text{ arcsecond (or } 0.8'')$$
$$\text{Optimal sampling} = 0.5 \times \text{Theoretical resolving power}$$
$$= 0.4 \text{ arcsecond per pixel}$$

To conclude, we have to find a way to cast the image on the sensor so that 0.4-arcsecond details correspond to distinct photosites by adjusting the resulting focal length.

6.1.1   General formula

Now it is time to calculate the resulting focal length in order to cast 0.4-arcsecond details on 5.6-µm photosites with the help of a simple formula:

$$\text{Optimal focal length} = 206 \times \text{Photosite size}/\text{Optimal sampling}$$
$$= 206 \times 5.6 \text{ µm}/0.4 \text{ arcsecond}$$
$$= 2884 \text{ mm}$$

In this example, we have a 150/750 telescope, that is a focal length of 750 mm, insufficient by far relative to the desired 2884 mm. We have to increase the resulting focal length, by applying magnification with the help of an eyepiece projection or a Barlow lens. With a 16-mm eyepiece and a distance of 80 mm between the sensor and the eyepiece, or by stacking two, 2× Barlow lenses, the resulting focal length can be made to reach 3000 mm, with a sampling rate of 0.38 arcsecond per pixel, which is very close to the optimal value. The resulting F/D ratio is 3000 mm/150 mm = 20. Adopting wider or smaller photosites leads to the need to adapt the resulting focal length (and the F/D ratio) to keep the same sampling rate.

6.1.2   Accurate formula

As we saw in Section 2.3, the resolving power depends on the telescope aperture and the wavelength (color). Given these parameters, the following formula gives the required resulting focal length for an optimal sampling rate:

$$\text{Resulting focal length in millimeters} = \frac{W}{\text{tangent}(S \times N \times 1.22 \times 10^6 \times \lambda/D)}$$

where

W = sensor width in millimeters;
S = oversampling: 0.3 to 0.5 (for 0.3–0.5 times the theoretical resolving power);
N = number of horizontal photosites;
λ = wavelength in nanometers (approximate values: blue = 470 nm,
    green = 520 nm, red = 650 nm);

D = diameter of the telescope in millimeters.

The tangent is calculated by using a calculator or a spreadsheet.

We assume that all modern sensors have square photosites.

6.1.3 Adjusting the sampling and the F/D ratio

In practice, the F/D ratio cannot be freely adjusted. Its limit depends on the instrument type, its optical quality, and its diameter. Here are some tested examples.

- A 150/1200 (6-in), apochromatic triplet refractor: a resulting F/D ratio of 37 has been successfully exploited.
- A 254-mm (10-in), F/D = 4.7 Newtonian: good results up to F/D = 24.
- A 90-mm (3.5-in), F/D = 13.8 Maksutov: excellent results at F/D = 13.8 prime focus.

The tests were performed with the same monochrome camera having 5.2-μm photosites. A camera with larger or smaller photosites would have led to different F/D ratios. The experiment only had the purpose of comparing the respective magnifications in comparable circumstances. For example, a resulting F/D = 45 is feasible with larger photosites, a color camera, and a 102-mm (4-in) Maksutov.

This reveals that the apochromatic triplet refractor has a magnification capability far higher than its optimal sampling rate. This is not a problem, because the contrast remains excellent even at very high magnifications. The Newtonian has a lesser optical quality but, given its greater diameter, the optimal F/D ratio is better related to the optimal sampling rate. The limit of the Maksutov is related not to its quality, which is very good, but to its small diameter.

Reflectors are penalized by the reflectivity of the mirrors: a loss in brightness of 6%–8% per mirror (1% for dielectric ones). In contrast, good refractors have excellent light transmission, often 98%–99%. Reflectors also require a higher accuracy in grinding and polishing than refractors.[2]

Another criterion is the prime-focus F/D ratio. The image of a star at the focal plane is larger if the F/D ratio is "slower," that is, an F/D = 5 telescope should be more adapted to imaging than an F/D = 10 telescope. The diameter of the diffraction spot, or Airy disk, of a star is given by

$$\text{Diffraction spot (nm)} = 2.44 \times \lambda \text{ (nm)} \times \text{Focal length (mm)}/\text{Diameter (mm)}$$

---

[2] Almost double that for a refractor to provide similar results for a given wavelength. This accuracy is rated in terms of the wavelength fraction, measured peak-to-valley (more objective) or RMS (root mean square, a kind of average). A good reflector requires a minimal optical accuracy of about λ/6 (one-sixth the wavelength of light); a refractor requires half that value.

For instance, an F/D = 8 telescope shows at prime focus a diameter of the diffraction spot of 10 700 nm, or 10.7 μm, twice the size of the diffraction spot of an F/D = 4 telescope, given that an average photosite of a planetary camera is 3.5–5.6 μm wide. Knowing that, a camera at prime focus should be sufficient to get the tiniest details of the cast image. This is the case with small-aperture, F/D = 15 Maksutov telescopes.

The sampling rate is related to the capabilities of the instrument. This is checked experimentally by adjusting the distance between the (non-telecentric) Barlow lens and the sensor, or the distance between the eyepiece and the sensor. If the atmosphere is steady, measuring the contrast and the effective resolution of images taken with different adjustments can reveal what the best combination is: this is a balance between an acceptable contrast (decreasing with magnification) and the best resulting resolution, when no more details are visible at higher magnification. The limit to the exploitation of an instrument will be encountered more quickly if the instrument has a small aperture, because the turbulence remains below the resolving power. It is better to perform tests on several nights per year because of seasonal, atmospheric variations. A 125-mm (5-in) telescope provides regular results almost all year long, whereas a 300-mm (12-in) telescope is very prone to turbulence variations.

The mean turbulence (specified in arcseconds) depends on the quality of the site:

- poor observing site: 1.5–2″
- average observing site: 1–1.5″
- very good site: 0.7″.

These values should be divided by 2–2.5 to find the appropriate sampling rate. This directly indicates the best suited instrument for a particular site with the help of Table 2.1. The perfect instrument is supposed to be the one with an optimal sampling rate in direct relationship with the turbulence. In practice, a 178-mm (7-in) refractor or a 300-mm (12-in) telescope achieves the possible resolving power at an average observation site. Since the contrast depends also on the diameter, and the turbulence sometimes ceases, we may also be attracted by the idea of using a larger telescope, even if the instrument can but rarely be exploited in appropriate conditions.

6.2 Colored and dichroic filters

Colored filters are mainly recommended for observing, but imaging is possible. They are cheap and reliable, being dyed in the mass and resistant to abrasion. The selectivity is not very accurate, in that the transmission is moderate and greatly exceeds the central bandpass with another, possibly undesired peak in the near infrared or near ultraviolet.

Here are some useful filters for lunar imaging with a good transmission:

- #38A (deep blue): the maximal transmission is 58.8% (total transmission for all colors is 17.3%);
- #11 (yellow–green): the maximal transmission is 60.2% (total 40.2%);
- #25 (deep red): the maximal transmission is 89.5% (total 14.0%).

If the transmission is too weak, leading to excessively short exposures, light filters may be a solution:

- #82A (light blue): the maximal transmission is 82.4% (total 72.5%);
- #23A (light red): the maximal transmission is 90.2% (total 25.0%).

Dichroic filters exploit the phenomenon of interference to pass only the desired band of colors. They are more expensive, but extremely selective, and the transmission is excellent, 95%–98% in the entire bandpass. Some dichroic filters include near-infrared and near-ultraviolet rejection. Lunar imaging accepts non-infrared/ultraviolet-rejecting filters, exception for achromatic refractors, which require an additional minus-violet filter: these instruments do not focus correctly at short wavelengths, while infrared wavelengths are treated correctly and not highly reflected by the Moon (in contrast to the case for Mars).

If you can afford dichroic filters, they are highly recommended, because they pass 8%–40% more light than colored filters, depending on the color, and because they are substantially more selective and hence more efficient. But simple #29 (transmission peak 90.4%) or #70 filters (80%) are good bandpass filters for the red and infrared.

6.3 The atmospheric-dispersion corrector (ADC)

Owing to atmospheric refraction, when the air acts like a prism, a slight shifting between blue and red images may appear (Figure 6.2). This alters the resolution when the Moon is relatively low. If the optics is a telelens or a refractor, this may be mixed with chromatism. But the difference is easy to detect: when one pole is red while the other is blue, this is due to refraction (Figure 6.3). If the whole limb is blue or green, this is chromatism. Pure reflectors show only refraction as long as the Barlow lens or the eyepiece is appropriately corrected. When we use a color camera, a simple splitting into the fundamental colors is a solution: the green layer is always the best. If the camera is monochrome, the differential effect of the refraction is no longer split into different colors: it is entirely mixed up and the image is blurred.

A remedy is to use a colored or dichroic filter to image the Moon through a narrow bandwidth, eliminating the effect of refraction at the price of an increase in the duration of exposures. Another solution is using an atmospheric dispersion corrector (ADC), a set of two orientatable prisms to counteract the natural prism of the atmosphere. The shift in colors is neutralized, resulting in a gain in

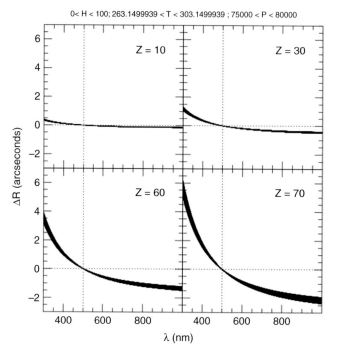

0< H < 100; 263.1499939 < T < 303.1499939 ; 75000 < P < 80000

Z = 10   Z = 30   Z = 60   Z = 70

ΔR (arcseconds)

λ (nm)

Figure 6.2 Measure of the atmospheric refraction. It depends on the air mass, humidity, and other factors. Z is the inclination angle relative to the zenith (in degrees). The blue is at around 450 nm and the red is about 650 nm. The deviation ΔR between the blue and the red is almost as much as 1.5 arseconds 60° away from the zenith, or 30° above the horizon. Diagram by ESO.

Figure 6.3 The atmospheric refraction shifts the colors: the limb at the North pole is bluish while the South pole is reddish. The dispersion increases when the Moon approaches the horizon. The image is cropped but unprocessed. In contrast, the chromatism of the optics does not depend on the height of the Moon and acts in the same way on the whole image: the limb is greenish or reddish everywhere. Unprocessed frame taken with a pure reflector and a DSLR.

brightness and accuracy. Some ADCs pass wavelengths in the near ultraviolet, some do not; this is irrelevant for lunar imaging (including petrographic imaging).

If we have neither a narrow-bandwidth filter nor an ADC, we can use a color camera with no filter. The red and blue components can be shifted afterwards with the help of the "RGB shift" functionality in RegiStax. This acts with an accuracy of one pixel only, but a preliminary, linear resizing (with no interpolation) of the image by 2× or 3× helps.

7

Essential image processing

7.1 Monitor calibration

The very first step of image processing is screen calibration. When we examine the same images on different screens, especially laptops, there are noticeable differences in colors and, most importantly for the Moon, in contrast. Laptops often show a certain brightness in low levels (dark areas on maria), while high levels (crater rims) are bright but similar values are mixed, resulting in an apparent overexposure. The same phenomenon sometimes occurs with video projectors: we may have to lower the contrast or the luminosity to reveal bright details. Desktop screens also show differences in contrast. The contrast of a screen is given as a ratio, for instance 1000:1 means that the difference between the lowest and highest levels is 1000. But 1000 what? Light is measured in cd/m^2, where cd is for candela (the candle remains a reference!). A monitor can emit the same amount of light as several hundred candles per square meter, for instance 400 cd/m^2. Since the Moon is very contrasted, and because most lunar images are monochrome, we have to carefully adjust the ability of the monitor to correctly display different light values, whether they are bright or dim, even if they are very similar (Figure 7.1). More importantly, we have to ensure that different monitors will show the same differences, more or less. That is why numerous websites display a monochrome calibration chart.

The untrained human eye can distinguish at least 100 different levels of brightness. Our lunar images practically contain at least 128 levels (a 7-bit coded image). Images from experienced astrophotographers contain several thousand different levels, up to some million shades. This dynamic range has to be reduced to match the limited contrast of the monitor. The starting point is to establish a ladder of significant values, especially the lowest and highest levels, corresponding respectively to a dark sea at the terminator and a bright, young crater rim. Then we have to select a number of intermediate shades. In practice, only 10–20 shades are necessary. Let us draw a rectangle composed of ten squares, each of them displaying a different shade. Each shade has to be chosen with respect to its coding light value rather than to the senses of the human eye. Most image editors have a special tool, generally called an eyedropper. It measures the color in terms of red, green, blue, or simply the level in the case of a monochrome image.

Numerous websites offer such a calibration chart, many of which are complex because they are intended to be used to perform various adjustments for color,

Figure 7.1 This chart contains sixteen shades, each of which corresponds to an evenly distributed coding value in 8 bits (256 possible shades divided by 16 makes sixteen quantified shades). The shades evenly decrease from 255 to 0, independently of the human eye's perception. Then the monitor has to be adjusted in contrast and lightness in order for all of the shades to be distinguishable. This calibration is quite simple, and is essential before we process and then share images.

sharpness, and image distortion for broadcast. For our purpose, the simplest is best. Once we have a chart, the monitor has to be adjusted in contrast and lightness to distinguish every shade on the chart. This can be done in several ways:

- by the OSD (on-screen display) menu of the monitor;
- by Intel® or NVIDIA utilities, "color management" with Linux Gnome, Apple → System Display → Color → Calibrate with OS X, or any other adjustment utility software provided with the computer, graphics card, or operating system;
- with the help of a calibration probe (e.g. datacolor Spyder 4/5, X-Rite i1Display Pro, and others).

Note that PhotoShop® offers resident software for permanent calibration (gamma) and an on-demand calibration (in the menu View → Proof set-up then Proof color).

7.2 Processing flowcharts

Basically, an image may be correct at first glance and we need do nothing but admire it. But experienced lunar astrophotographers can transform a good image into a beautiful image. The better the original image, the more powerful the enhancement. The following flowcharts describe different typical cases of processing. The detailed functions are presented in the next chapters. Of course, many variations are possible.

Let's start with the simplest equipment, for instance a smartphone in digiscopy (held by hand behind the telescope) and a single frame (Figure 7.2).

Now it is time to take the leap of a more efficient image enhancement. This requires a movie, or a large set of successive frames, to increase the signal/noise ratio and minimize the effect of the turbulence. Thus, more powerful processes can be applied, which otherwise could severely damage a single image (Figure 7.3).

Sometimes, a process results in a well-balanced image, with a high contrast between seas and highlands or young craters, but details remain unclear. If we try to increase the sharpness, the image becomes noisy. On the other hand,

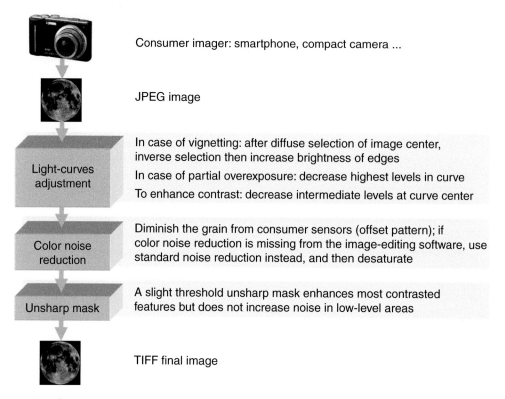

Consumer imager: smartphone, compact camera ...

JPEG image

Light-curves adjustment
In case of vignetting: after diffuse selection of image center, inverse selection then increase brightness of edges

In case of partial overexposure: decrease highest levels in curve

To enhance contrast: decrease intermediate levels at curve center

Color noise reduction
Diminish the grain from consumer sensors (offset pattern); if color noise reduction is missing from the image-editing software, use standard noise reduction instead, and then desaturate

Unsharp mask
A slight threshold unsharp mask enhances most contrasted features but does not increase noise in low-level areas

TIFF final image

Figure 7.2 A simple process to enhance a single frame. See an example with a smartphone in Figure 1.5. Since the input image is compressed (JPEG), with a noticeable amount of noise from a cheap sensor, little improvement is to be expected, but the output image is better balanced and the details seem more accurate.

a detailed image may lack contrast, and increasing the contrast leads to increased noise. We can simply combine the best of the two possible results by fusing them after parallel processes, one for the contrast, one for the accuracy (Figure 7.4).

These three flowcharts are summaries. Furthermore, each type of lunar feature has to be processed in a particular way (e.g. craterlets, domes ...), and, sometimes, each individual process may be applied iteratively with various parameters to unveil details at different scales (e.g. ray systems). We will see in the next chapters how to easily realize each step for various kinds of lunar features.

7.3 Stacking

Even if the Moon is very bright, high magnifications and peculiar subjects like earthshine or lunar domes cause images to be noisy. Subsequent sharpening

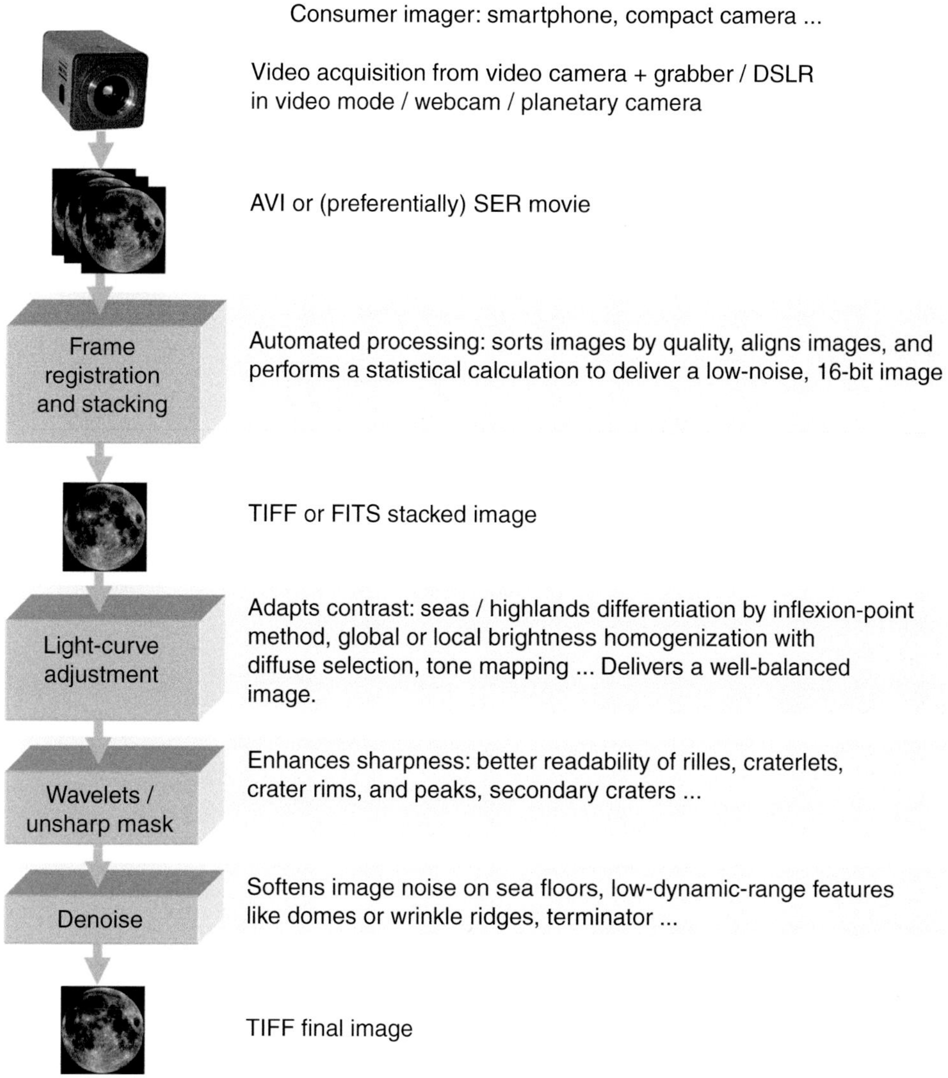

Consumer imager: smartphone, compact camera ...

Video acquisition from video camera + grabber / DSLR in video mode / webcam / planetary camera

AVI or (preferentially) SER movie

Frame registration and stacking

Automated processing: sorts images by quality, aligns images, and performs a statistical calculation to deliver a low-noise, 16-bit image

TIFF or FITS stacked image

Light-curve adjustment

Adapts contrast: seas / highlands differentiation by inflexion-point method, global or local brightness homogenization with diffuse selection, tone mapping ... Delivers a well-balanced image.

Wavelets / unsharp mask

Enhances sharpness: better readability of rilles, craterlets, crater rims, and peaks, secondary craters ...

Denoise

Softens image noise on sea floors, low-dynamic-range features like domes or wrinkle ridges, terminator ...

TIFF final image

Figure 7.3 A two-step process to get a much better image. The movie (or a large number of uncompressed frames) is acquired by the camera. The stacking step is not only to deliver a high signal/noise ratio: in addition, frames are automatically (or manually, if we are persevering) sorted and weighted according to their accuracy (local contrasts). The intermediate, stacked image is viewable, but it seems blurred. Indeed, it has a very high number of distinct levels that the human eye is not able to distinguish. The second step is to emphasize the details.

processing will increase both details and noise, so frames must be as clean as possible. The remedy is stacking frames to calculate a kind of average image, or performing some other statistical calculation, such as of the median or kappa–sigma. The improvement is expressed as

201

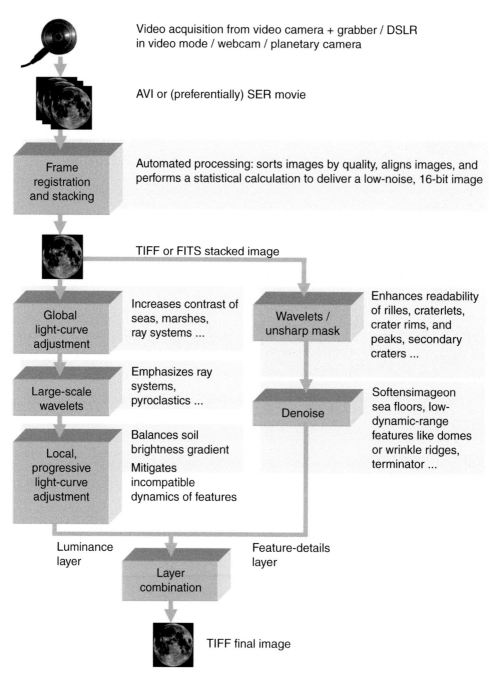

Figure 7.4 The stacked image is duplicated and transformed as layers. Each layer is processed separately. The first one is processed for contrast only, regardless of the accuracy (it becomes the "albedo layer"). Then the first layer is hidden and the second layer is processed for sharpness only, regardless of the contrast (the "features details layer"). The fusion consists of the combination of the two layers using a particular mode depending on the software: "multiply," "luminance," "hard light," "screen" ... The layers may be swapped and adjusted in terms of percentage transparency to vary the result. The parallel processing is an efficient method to get contrasted and sharp images with little noise.

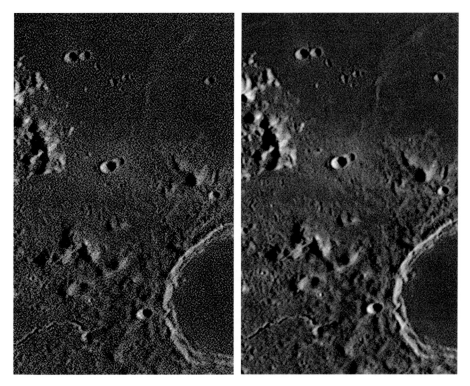

Figure 7.5 Left: 146 images stacked. Right: 600 images stacked. The processing was rigorously identical and the frames come from the same movie (5° W, 49° N).

$$\text{Signal/noise ratio} = \sqrt{\text{Number of frames}}$$

In other words, 25 frames give 5 times less noise, 100 frames give 10 times less noise, and so on (Figure 7.5). In practice, the improvement limit is quickly reached, and using too many frames will not be useful because of variations in turbulence, thus some 200–500 frames should be a good compromise. Using more frames does not result in an obviously better signal/noise ratio. Some astrophotographers take thousands of frames in poor conditions, in order to keep some hundreds of good frames.

This sounds a little complicated, but software devoted to planetary (and deep-sky) imaging helps a lot with automated processing. We only have to load the movie (AVI or SER), click on some buttons, and the software calculates the stacked image. The best types of free software for stacking are AutoStakkert! 2 by Emil Kraaikamp, RegiStax by Cor Berrevoets, AviStack by Michael Theusner, IRIS by Christian Buil (the latter includes numerous, brilliant features but it is not user-friendly for beginners), and Lynkeos for OS X. Commercial software packages are very efficient too, including many other functionalities, e.g. AstroArt, Autopano Pro, MaximDL (mostly used for deep-sky imaging), and Astrostack. In practice, the first three freeware programs are the most widely used for stacking.

Stacking software can open SER (the best choice) or AVI movies, or still images. The resulting, stacked image has to be saved in FITS or TIFF format.

- In the main window of AutoStakkert! 2, one has to select "Surface" in the first tab and click on "Analyze," then, in the image window, "Place AP in Grid" (AP stands for alignment points, to re-align the local areas of the image which have been distorted and scattered by turbulence). The AP width may be placed manually and enlarged in the case of highly magnified images, where details are less contrasted and not as abundant, set "Min Bright" to 0. One may choose different values for "Frame percentage" depending on the average image quality, but the default value of 50% is often sufficient. Finally, the "Stack" button completes the stacking process.
- Avistack also has to be set in "Surface" mode: this is the "Alignment Type" tool in the "Frame Alignment" folder located in the "Parameters and Settings" window. Avistack is the most automated, thanks to its "Batch processing" button. It is recommended for beginners.
- With RegiStax, successively click on "Set Align points," then on "Align," "Limit," then "Stack." Since Avistack and AutoStakkert! 2 usually provide better results, a wise strategy is to use one of them for stacking, and then process the image with the remarkable wavelets of RegiStax.

Once the movie frames have been stacked, the parameters are implicitly saved, so they can be used for further or batch processing. Avistack can load several movies at one time in the "File → load" menu, while with AutoStakkert! 2 the user just needs to drag and drop the list of movies on its main window.

What are the advantages of stacking?

- Lowering noise: a median calculation based on some hundreds of images greatly lowers noise.
- Frames taken during worst seeing can be eliminated, keeping the best frames. Moreover, the software (or the astrophotographer) can estimate the quality of each frame and weight each frame according to its sharpness (e.g. by estimating local contrasts); thus the best frames have more importance in calculating the stacked image than blurred frames.
- The calculations are performed with high accuracy, that is in 16 or 32 bits (sometimes even better with "floating-point arithmetic," a method for a computer to calculate a value with numerous decimal places if needed).
- Each frame is broken down into numerous small areas, and the software aligns similar parts to lower the scattering caused by turbulence. This is done by assuming that highly contrasted, easily recognizable, small features such as

craterlets are the same features moving away from an average location in different frames. The different areas are repositioned at an average location to compensate for the distortion by turbulence. This also takes into account drifting during movie acquisition (e.g. if the polar alignment of the telescope was flawed, or drifting caused by windy conditions).

• Stacking is totally or partially automated.

What are the drawbacks of stacking?

Some software packages are not suited to stacking large images. At the time of writing, Autoskakkert! 2 is the best performing, especially because it manages multipoint registration very well, that is, image breakdown before repositioning each area accurately – a fully automated feature, but registration points can be manually positioned. Avistack is the most convenient for beginners because it is totally automated. The author usually uses AutoStakkert! 2 or Avistack to load several tens of SER movies prior to starting stacking in batch mode. The software automatically stacks the movies throughout the night, and FITS images are ready for post-processing (sharpness, contrast …) the next day. Multipoint-registration stacking demands a powerful computer – or time to wait. Smaller images, such as images in VGA format (640 × 480 pixels) from a webcam or a DSLR are more quickly stacked than large images from megapixel planetary cameras. In most cases, AutoStakkert! 2 and Avistack produce very similar results.

7.4 Some traps to avoid in lunar image processing

Before we get into lunar image processing in depth, we have to know about some common image alterations. They occur because of an inappropriate choice of image-editing software (e.g. using internal, 8-bit processing in place of 16 or 32 bits leads to posterization), the use of compressed image file formats (the accuracy is reduced and artifacts appear) in place of uncompressed formats such as TIFF or FITS, the use of inappropriate or immoderate noise-attenuation or sharpening processing (the image becomes blurred or noisy with artifacts), or because of image resizing (Figure 7.6).

The best advice is to always have a backup of the stacked image, moderately process the image to be kept as the reference image, and then try different and successive processes while having a glance at the reference image from time to time. A long processing session may result in a final image that is worse than the reference image! In such a case, it is a good idea to suspend the processing for some days before trying other algorithms or workflows. After gaining experience, a new processing workflow may reveal itself to be better suited, encouraging the re-processing of older images with better results.

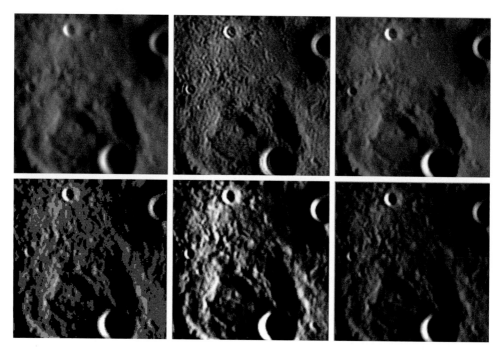

Figure 7.6 Some classic traps to avoid while processing an image. Top left: the stacked image (extremely cropped!). Center left: an excessive unsharp mask results in imaginary details, rough surfaces, artificial circles in craters, and outlining of inner surfaces. Top right: a median-type denoising wipes out details. Bottom left: posterization by reducing the number of levels, or expanding adjacent levels, creates abrupt transitions of shades. Bottom center: an exaggerated contrast expansion clips high levels (internal rims and scarps). Bottom right: decreasing low levels conceals noise while it increases the readability of details, but the whole image is darkened (15° E, 11° S).

7.5 Sharpness

7.5.1 Acutance and unsharp mask

Acutance is a quasi-magical legacy of film photography. It consists of artificially enhancing sharpness by locally lowering low levels close to high levels and vice versa where a great difference in brightness occurs on the film. In other words, this increases local contrasts. This was further extensively exploited in analog video imaging. Nowadays, the expression "unsharp mask" (another technique from film photography) is used in place of acutance, likely because the results are very similar if the technique is applied on the edges of contrasted areas of the image. It remains the basic way to improve the readability of a poorly contrasted image. But nothing is perfect, and applying an unsharp mask increases noise at the same time (Figure 7.7). Small-scale contrasts define the image resolution, and, if the

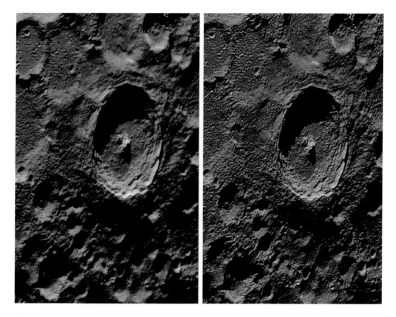

Figure 7.7 Applying an unmoderated unsharp mask enhances the readability of details along with the noise, creating artificial patterns (11° W, 43° S).

sampling is right, this scale is similar to that of the pixels, and hence to that of the noise. A strong sharpness enhancement generates artificial patterns that can be confused with real details. As with any special effect, it must be used sparingly, preferably with high-signal/noise-ratio images.

The two essential parameters of unsharp mask are as follows.

- Width, or radius: the number (integer or decimal) of pixels around areas whose edges show strong, local contrast; in other terms this represents the accuracy of local contrast enhancement. The width has to be commensurate with the original resolution of the image in order to produce a natural appearance. In most cases, a width of 0.7–2.0 pixels is convenient.
- Efficiency, or strength: the efficiency (specified as a percentage) of the effect. It has to be adjusted to the image with respect to how much the noise increases.

Some software packages offer advanced unsharp-mask processing with additional parameters, such as a threshold level, to protect low levels against noise augmentation, or a progressive limit to preserve high levels because they are supposed to be more accurate than low levels. For instance, Photoshop® proposes a standard "unsharp mask" and a parametric "smart sharpen," and Fitswork – a freeware program originally devoted to deep-sky image processing, by J. Dierks – offers an adjustable "Gaussian sharpening" and a more classic

"Unsharp masking" (in the menu Processing → Special filter → Unsharp masking).

7.5.2 Multiple-resolution unsharp mask

Despite being simple to use, unsharp masking may be exploited in a less conventional way. Since a lunar image contains many scales of features, from large maria with little contrast to highly contrasted but narrow rimae, unsharp masking may be applied in several, successive steps with various widths and strengths. In other words, we can apply multiple-resolution unsharp masks. The sole limit is imposed by the signal/noise ratio, because successive unsharp masks greatly increase noise. If the image comes from a stacking of some hundreds of frames taken with low gain, the noise will remain moderate. The author successfully applies successive unsharp masks with Fitswork with the following parameters:
Processing → sharpen filter → Gaussian sharpening

- Step 1: Radius = 10.0; Threshold = 0; Strength = 150
- Step 2: Radius = 1.3; Threshold = 0; Strength = 350
- Step 3: Radius = 0.6; Threshold = 0; Strength = 350

The first step is to greatly enhance the contrast of moderately large areas, especially crater rays. The last two are accurately adjusted to the finest details of the image: crater rims and central peaks, craterlets, and rimae. Two important points are that one should not apply the same width more than once, since doing so will overwhelmingly increase the noise, and one should keep moderate values for strength. In the previous example, the maximum strength is 500, and the first step is applied with a very moderate strength to preserve the smoothness of the originally noiseless image. The same principle may be applied with numerous image editors like GIMP.

7.5.3 Wavelets

Wavelets are formal representations of quickly evolving phenomena. They can be understood as the next step to the fast Fourier transform (FFT), that is, a breakdown of a signal into its basic components, namely a set of frequencies and amplitudes. Wavelets can formalize more complex signals, especially if there exists a supplementary dimension, such as time, or if the signal is an image rather than a sound, that is, a two-dimensional, quickly evolving signal (e.g. a crater with a complex shape). If we have already tried to process an image with a multiple-resolution unsharp mask, wavelets will not seem totally strange to us: imagine that we could simultaneously apply a set of parametric unsharp masks. Wavelets are one of the most powerful – and popular – sharpness enhancements. Numerous astronomical image-processing software packages offer this functionality, but the most popular is RegiStax 6 by Cor Berrevoets's team, which

Figure 7.8 Wavelets increase contrast at different scales with Cor Berrevoets's RegiStax 6. Top: original image. Center: levels 1 and 2 reveal the finest details – which are often mixed with noise – such as craterlets and rilles. Bottom: levels 3 to 4 increase the contrast of large-scale features, such as rims and wrinkle ridges. There are six levels with adjustable scales and denoising (1° W, 29° S).

features some extremely valuable parameters, including noise killing (Figure 7.8). The freeware is well documented (www.astronomie.be/registax/ previewv6-3.html), but we can have a first look at essential parameters for lunar images.

- Once the previously stacked image has been loaded, we have to choose the wavelets scheme, either *dyadic* or *linear*. The first variant, which is extremely efficient, is sometimes used with low-contrast planets like Venus and Jupiter. Since the Moon is very contrasted, the linear scheme is more suitable.

- *Initial layer* and *Step increment* defines the starting point, which is related to the resolution of the image, and the progression interval, related to the dimensional range of the features of the image. Default values are perfect for lunar images, though undersampled images may require an initial step of 2, and/or a greater step increment.
- *Use Linked Wavelets Layers* affects the processing by chaining the six layers rather than working with six separate layers, improving the efficiency. This also increases noise, and this functionality cannot be correctly used without incrementing the *Denoise* value.
- The main task is the setting of each layer's parameters.
  - The numbered *checkbox* activates the layer.
  - Increasing *Denoise* lowers noise at the price of a weaker wavelet efficiency for the layer which is being considered.
  - *Sharpen* is the efficiency of layer sharpening. We have to find the right balance between efficiency and denoising. Generally a smooth progression is sufficient, for instance Denoise is 0.15 for the first layer (the most noisy except for undersampled images), 0.10 for the second layer, 0.5 for the third layer, and 0.0 for the three last layers. Sharpness may be set to a moderate value for the first layer, e.g. 20, because it contains most of the noise (if the image was not undersampled). The second and third layers may be set to higher values, e.g. 50–80, and the last layers may be set to smaller values, e.g. 30, 20, 10, because they contain large-dimension features that do not need to be increased in sharpness. This depends intimately on sampling and turbulence, which affect the dimensions of details.

The whole set of parameters can be tuned according to typical shooting parameters and turbulence, then saved and reused later if images are taken in identical conditions.

7.5.4 Convolution and deconvolution

A convolution is basically the application of a mathematical function onto another function. The first one can be a simple arithmetic calculation to be applied by means of an array (a matrix) of numbers, representing a central point and neighboring points, to the image. This is widely used in image editors as "filters," for instance "low-pass" filters to smooth the image, or, inversely, a "high-pass" filter to increase contrast. We can even freely design a filter matrix in some image editors (e.g. with PaintShopPro™, or Photoshop® with menu Filter → Other → Custom). A very interesting filter is the deconvolution (applying a reverse function) provided by Fitswork in the menu Processing → Sharpen Filter → Deconvolution. A sharpening filter has to be applied prior to deconvolving the image. This greatly helps to reduce noise in an image, assuming that a flaw in the image is homogeneous and comes from the whole optical and imaging system. The author tried it successfully on binary stars and jovian moons to compensate for astigmatism of the mirrors. As stellar objects are supposed to resemble fuzzy points (a Gaussian distribution, in the form of a bell), any disturbance from the optics distorts these fuzzy points and results in curved shapes, like miniature comets, or commas. Taking a sample of the distortion by selecting a small rectangle containing the misshaped star amounts to creating the convolution matrix. It can be reversed by the software (as the "PSF image"), and then applied to the whole image, canceling the distortion and resulting in perfectly rounded stars. Of course, an image of the Moon shows irregular patterns rather than simple points, but this can help to attenuate optical flaws. More interestingly, the software proposes a default pattern, and, if we let it set default values for all parameters, then just set "image noise" to 0.5–2.0 and "radius" to 1.0–2.0, the filter applies the default deconvolution pattern to the whole image, cleverly fuzzying out noise without altering significant details of the image. The balance between "image noise" and "radius" is critical. The final accuracy of a sharp and moderately noisy image may be improved by a factor of 2 or more.

7.6 Contrast and inflexion points

Contrast is the most elementary adjustment of a lunar image. It is defined, after Michelson, as

$$\text{Relative contrast} = \frac{\text{luminance max} - \text{luminance min}}{\text{luminance max} + \text{luminance min}}$$

The luminance max and min refer to the brightest and darkest areas of an image. A contrasted image is easier to appreciate because lunar features appear more distinctly. Even with the cheap camera of a smartphone, the original image often contains subtle differences in contrast: details that our eyes cannot distinguish.

The debate about enhanced or faithful images when adjusting contrast is not very important since, from the very beginning of photography, numerous altera-tions of contrast (gamma, paper grade, multigrade filters, masking ...) have been used either to improve the visibility of certain details for technical illustrations or to enhance the mood of an image for artistic purposes. In the first case, the goal is to reveal topographic details and soil composition. In the second case, the goal is to render a particular feeling because the scene was inspiring (e.g. the Moon in clouds or in a landscape). Another problem is that the contrast of the final medium (photographic paper, a review) is never the same as that of the original medium (a digital image seen on a monitor). In all cases, some interpretation from the photographer is inevitable. Moreover, it attracts interest because the photo-grapher's style can assert itself and become mature. The most conspicuous enhancement to be expected is the differentiation of maria and highlands, pyroclastic deposits, and some details like the ray systems of Tycho or the darkness of the floor of Plato.

Some frequent terms occur in image editors. They often come from film photography, since some of the algorithms of image-editing software mimic (and greatly improve on) chemical and film combination processes. Here are the most frequent.

- Contrast: the difference between the lightest area and the darkest area in the image. Decreasing the contrast means that areas with the lowest levels of light are lightened while lighter areas are dimmed. Conversely, increasing the contrast amplifies noise in the areas with low light levels while all of the strongly lit parts of the image are mixed up. Adjusting the contrast by using free curves is more flexible.
- Brightness: a value added to all levels. Increasing the brightness helps to reveal the details of a dark image but it may clip the brightest areas.
- Gamma: an S-shaped curve expressing how extreme levels in the original image are translated into the resulting image. Once again, adjusting the contrast by using free curves is more flexible.
- Levels are the floor and the ceiling of the whole set of levels in an image. Increasing the floor value simply clips low levels: this can help to increase the contrast of an image, but this operation leads to posterization.
- Curves is a common tool to adjust levels more precisely.

If the original images show too little contrast, this may be a result of different causes: dew, bad collimation, too strong a magnification, inadequate optical quality, internal reflexions … But the contrast depends also on the phase. The lunar contrast varies greatly because, when a region is illuminated at right angles (the center of the Moon at full Moon), predominantly the soil composition and age affect the albedo. With low-angle illumination, topographic details cast shadows and very-low-altitude or shallow features are emphasized. The intrinsic variation of albedo is inevitable and does not depend on the photographer's interpretation. This provides a way to study the Moon's surface more accurately, in depth, with appreciable variations in both geological and aesthetic rendering.

An efficient and simple method to emphasize different regions of the lunar soil, as long as the illumination angle is high, is by determining an inflexion point (or two) in the curve. Figure 7.9 contains a part of a mare and a highland. At a first glance, we mainly distinguish the mare and the highland. But if we decrease low levels, then increase high levels, and finally slide a point of the curve at half the maximum level (Figure 7.10), not only does the difference between the two areas suddenly become obvious, but in addition an intermediate area appears, like some dark grains scattered in the highland. Once the right inflexion point has been discovered, its level is lowered, and then three areas of different albedo become differentiated.

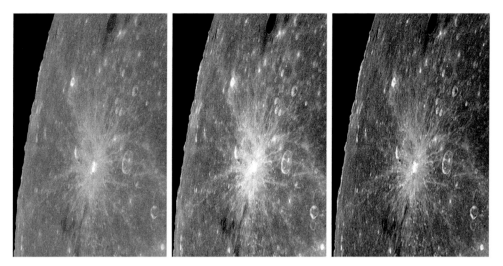

Figure 7.9 The bright Byrgius A crater and its ray system, topped by the dark Crüger crater. Left: the image is processed for sharpness only. Center: basic contrast adjustment with an image editor. The ray system of Byrgius A is emphasized, but all bright craters are clipped, while subtle shades in maria are mixed up. Right: contrast adjustment with the inflexion-point method. High levels are maintained (bright ejecta around Byrgius A) while low levels are strongly dimmed to reveal Mare Humorum and dark, flooded craters such as Crüger. The inflexion point was chosen to give highlands an intermediate level with a moderately steep slope in the curve. This greatly helps to differentiate intermediate-albedo terrains (64° W, 24° S).

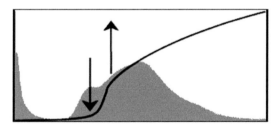

Figure 7.10 Contrast adjustment by means of a couple of inflexion points. This abrupt accentuation, which is determined empirically for each image, increases the contrast between maria and highlands and preserves bright craters from clipping. On horizontally sliding down the abrupt slope, the soils become strongly differentiated, subject to the condition that the brightness of the image is homogeneous for a given nature of soil, that is, far from the terminator. If the image contains a strong gradient, that is, a small colongitude, gradient compensation (Section 8.1) can be performed prior to determining the appropriate inflexion point.

We have spoken only of global contrast adjustment, but sometimes different areas of the image cannot be adjusted with the same curve, especially with a large-field image. We will see in Section 8.2 how to split contrast adjustment in different parts of the image.

7.7 Noise reduction … and how to avoid noise

Once the image has been processed for sharpness and contrast, we may detect false details, a kind of granularity or small artificial patterns, particularly on maria and other dark areas because the signal/noise ratio is lower. When we begin to use sharpness processing, there is a great temptation to exaggerate details. A number of lunar – and planetary – photographers obtain detailed but noisy images, where there is some confusion between authentic details and artifacts. The darkest areas of the image resemble JPEG images with excessive compression. The worst results are in low-light areas: artifacts, posterization, and remnants of the fixed-pattern noise are mixed up with genuine details.

The preventive measures one can take are as follows.

- Stacking more frames.
- Using 10-bit or 12-bit converters, but this solution requires a higher volume of data, more computing resources, and fewer frames per second; anyway, stacking fixes a lack of accuracy in analog-to-digital conversion.
- Decreasing the gain and increasing the exposure time while shooting.

The corrective action is as follows.

- Concealing noise with the help of a number of algorithms (Figure 7.11).

There are intermediate solutions.

- Using threshold levels for unsharp-mask functions.
- Contrast compression of low levels (Figure 7.12).
- Multiple-layer processing.

7.7.1 Avoiding noise during acquisition

Analog-to-digital conversion is less accurate in low-level quantization because it relies on too few bits, and because low levels are often close to noise. Some cameras actually offer reliable behavior in 6.5 or 7 bits (statistically) rather than 8 bits – the same phenomenon occurs with more accurate 12-bit or 16-bit converters, but the error is reduced if we consider the relative contribution to the set of levels.

The possible remedies are simple.

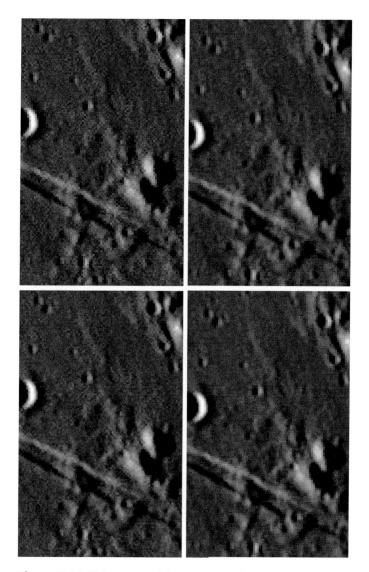

Figure 7.11 Using some of the numerous denoising algorithms. Top left: original image, strongly cropped. Top right: negative wavelets in RegiStax was set to +6.1 for level 1, −3.8 for level 2, and −3.2 for level 3. Bottom left: deconvolution in Fitswork with noise = 4, radius = 1.2. Bottom right: filter − denoise in Photoshop® with strength = 1, preserve details = 3%, sharpen details = 3% (13° E, 7° S).

- The fixed-pattern noise (FPN) can be subtracted from frames prior to stacking. This cannot be done to the stacked image afterwards because, during the stacking process, frames are aligned with respect to lunar craters, dancing with the turbulence, and because of mechanical imprecisions during tracking, while the

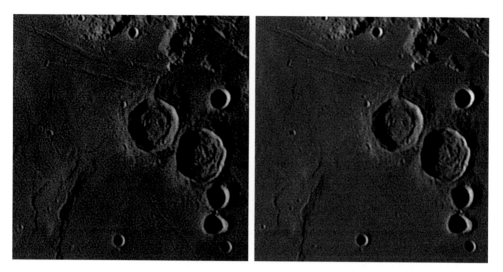

Figure 7.12 Left: the original image. Right: compressing the dynamic by leveling up the lowest levels while darkening the brightest ones; the noise is reduced with little alteration of the accuracy and no interpolation (20° E, 2° N).

FPN stays at the same place: the result is an addition of multiple instanciations of the FPN affected by random drifting. Some acquisition software (e.g. FireCapture) is able to acquire a series of dark frames – when the telescope is masked – to be automatically subtracted during frame acquisition. Although "hot pixels" in the dark frames depend on exposure and temperature while the FPN is constant, we can roughly consider that on-the-fly dark subtraction is sufficient for lunar imaging. Some sensors like the MT9M001 (which is nowadays deprecated, but was an acclaimed sensor of its time for large lunar images) have a large amount of FPN and random noise, but more recent sensors like the MT9M034 or the IMX174 show little FPN and little noise. Nowadays, the FPN, random noise, and hot pixels have become negligible for moderately short exposures, e.g. 15 ms or more, but the FPN remains obvious, intrusive, and unfixable with very short, 200-µs exposures and low gain. CMOS sensors require a neutral-density filter to dim the lunar brightness and to use longer exposures, but, curiously, the FPN remains present and hardly fixable if the acquisition software does not automatically subtract the FPN (this problem is present in solar imaging, too).

- Stacking more frames. Once again, the randomly drifted FPN is more or less spread in the stacked image. Just like for deep-sky images with an interlaced sensor, the artificial pattern is averaged with a median stacking and drowned in the background noise. Stacking some hundreds of images is enough to wipe out a great part of the noise. However, lines and columns still appear at the edges, because these parts of the frames are less frequently superimposed (in cases of improper tracking, turbulence, shocks, gusts . . .).

- Avoiding a strong gain, which is an artifical amplification of signal and noise. A longer exposure is preferable, and gain is adjustable in the final image with subtler algorithms than a simple multiplication. Moreover, datasheets show, for some sensors, a relatively complex built-in algorithm with difficult-to-manage, steep amplifications from certain levels. Tests have to be performed to determine the maximal gain with acceptable noise (e.g. 60/100 for an MT9M034, 150/600 with an IMX224).
- Some random noise originates in electric and electromagnetic phenomema (Section 2.9.11) and not-so-rare cosmic rays (appearing as transient, white dots).

7.7.2 Negative wavelets

We met wavelets in Section 7.5.3. But, with a bit of curiosity, we can see that the efficiency adjustment for each layer can move slightly to negative values. This is the case with RegiStax, Fitswork, and others. This means that noise can be lowered in different layers, or scales. This is a convenient fixing method when we have already applied too strong a sharpness process. A global blurring, for instance Gaussian blur (a classic functionality in numerous image editors), could cause a loss in detail at all scales. But noise is preferentially nested in fine details, and the possibility of choosing the noise scale, with the help of the multiple-layer calculation of wavelets, allows the softening of altered details at a precise scale. The operation simply consists of loading the noisy image, and setting the first and the second layers (where noise is most present) to slightly negative values. Each negative wavelet may be combined with additional denoise/blurring parameters (RegiStax, Fitswork). An example is shown in Figure 7.11.

7.7.3 Other noise-killer and image-restoration algorithms

Noise killing is an important field of research, and some commercial software programs are totally devoted to it. Many algorithms exist, and astrophotographers have spent a lot of time comparing their efficiency on astronomical images. Noise killing in astronomical images is very demanding because details must be preserved and, in addition, artifacts from denoising must be avoided. For instance, one of the most classic and destructive denoising algorithms is the *median* filter: noise is suppressed and replaced by irregular tiles. Blurring filters are too destructive for details. Better algorithms use deconvolution, which tends to iteratively carry out the reverse process of a point-spread function or any invariable alteration of details like turbulence, and, in some cases, optical flaws like astigmatism. In astronomy, Van Cittert deconvolution and Richardson–Lucy deconvolution are often used (e.g. in Maxim DL: filter → deconvolve; in Fitswork: processing → sharpen filter → deconvolution; in IRIS: by command line; in Photoshop: as plugins from Astra images and others, and so on). Other algorithms exist,

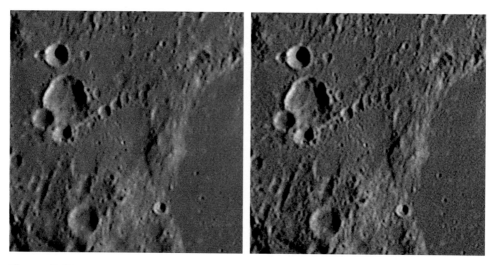

Figure 7.13 Left: a classic unsharp mask. The global sharpness is increased along with noise in low-level areas. Right: the threshold level has been adjusted to prevent low-level areas from undergoing processing – and an increase in noise – while high levels are sharpened (1° E, 8° S).

including least squares, maximum entropy, and Werner deconvolution. The primary goal is to reveal details without increasing noise, especially when restoring images. But there is another use of these algorithms: noise suppression. For instance, a side-effect of deconvolution is the noise reduction with or without detail enhancement. An example is shown in Figure 7.11.

7.7.4 Threshold unsharp mask

This is a preventive process to reveal details without increasing noise. In lunar images, low-level areas like maria are close to noise, whereas light areas such as ray systems and highlands have a more favorable signal/noise ratio. The idea is to enhance the sharpness of the high-level areas alone (Figure 7.13). This is done by applying the threshold parameter of unsharp mask: it determines the minimal level at which to apply the function, e.g. in Photoshop (filter → unsharp mask) or Fitswork (processing → sharpen filter → Gaussian sharpening). A different strategy is provided by the advanced filter → smart sharpen of Photoshop: the efficiency of the sharpening is mitigated by two sets of fading parameters, for low and high levels. Thus, the unsharp mask may be applied only to areas with intermediate signal/noise ratios, without clipping and fusing bright areas such as internal rims and scarps.

8
Advanced image processing

8.1 Balancing the terminator gradient

Near the terminator, the spherical lunar surface gradually passes into darkness. The image is unbalanced and numerous shallow reliefs are just visible. Moreover, such an unbalanced image is poorly suited to publication in a magazine, which requires pronounced differences in levels between dark and bright areas. A simple increase of the levels in dark areas is feasible with a diffuse selection at the price of increased noise and posterization. Too perfect a balance is not desirable because it could lead to a loss of the three-dimensional appearance of the surface: the Moon really is spherical. This delicate balance adjustment is greatly simplified with some special functions:

- Photoshop®: Image → adjustment → Shadow/Highlight (Figure 8.1)
- GIMP: colors → Retinex

If our favorite software does not offer a similar function, a diffuse selection (Section 8.2) may greatly help to reasonably increase the brightness: the very lowest levels should remain unaltered (to avoid posterization), while the selected area can be brightened.

Although the effect is understated, such a tool is invaluable to unveil shallow reliefs, domes, wrinkle ridges, and reliefs of less than some hundreds of meters which cannot cast a conspicuous shadow. The trap to avoid is exaggerating the application of the gradient correction, which could lead to an apparently planar lunar surface.

8.2 Local contrast with diffuse selection

We saw in Section 7.6 how to improve the readability of an image with global contrast adjustment. However, different areas of the image may contain different local contrasts. For instance, a mare or discrete dorsum has a limited dynamic range, and the contrast cannot be expanded without increasing noise and posterization. On the same image, another feature like a young crater may have a strong dynamic range and the signal/noise ratio is high: the contrast may be expanded without increasing the noise (Figure 8.2). Any contrast adjustment applied to the whole image would cause uneven results.

Unlike the two-layer processing, which acts on the full image with two different processes, we have to apply a process on different parts of the image. Generally no

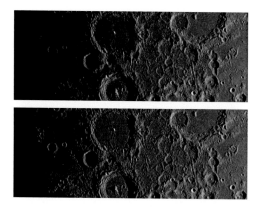

Figure 8.1 Top: wrinkle ridges and floors of large craters gradually disappear in darkness. Just the "Shadows" parameters are used (Amount = 31; Tonal Width = 0%; Radius = 5 pixels). Bottom: the correction mitigates the gradient and reveals dark reliefs with very little noise and hardly any posterization, while brighter areas remain unchanged. The effect is subtle but seriously improves the image's readability (0° E, 15° S).

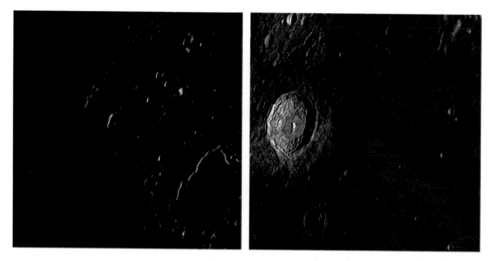

Figure 8.2 Left: the darkest part of an image, close to the terminator, contains mostly poorly differentiated, low levels. Since expanding the contrast increases noise and posterization, this part has to be processed gently. Right: the high-dynamic-range part with intermediate and high levels may be more strongly processed. A simple, global contrast adjustment is not suited to the whole image (centered on 49° W, 25° N).

more than two distinct regions are involved, but we can imagine that three or more regions could be processed separately. To select a part of the image, we use the classic "selection" tool called "Lasso" in Photoshop® and PaintShopPro™ and "Free select tool" in GIMP.

Here are some examples of how to do a diffuse selection.

• Firstly, we select the "Lasso" tool. It may have a "feather" or "progressive" parameter in the tool dialog box, but it is preferable to set it to zero initially. With the lasso, we roughly draw around the desired area, whatever its shape.

Then we adjust the progressiveness of the selection (Photoshop® and GIMP: select → feather; PaintShopPro: selection → modify → progressiveness). A value of 100 is a good starting point with a one-megapixel image.

- Then we adjust the selected area in terms of contrast.
- The same manipulation may be carried out for another part of the image with a different contrast value.

Note: some software packages limit the progressive selection area to the boundaries of the image, and the edges of the area are not included. The remedy is to enlarge the canvas – not the image but its frame – so that the diffuse area perfectly covers the desired part of the image.

The feathering must not be too steep. A value of about 20% or 30% of the image matches most cases, because the transition is large and soft enough to appear seamless. The drawback of a large feathering of local contrast adjustment is that it doesn't preserve the apparent sphericity of the Moon. It is better to let the soil keep a certain gradient due to the Moon's three-dimensional shape. Near the terminator, the curvature is perceptible over distances as small as 150 km (93 miles), representing, for instance, 20% of the image.

Figure 8.3 shows a contrast enhancement achieved by increasing the brightness of the large-signal/noise-ratio part of the image. The trick is also useful to diminish the darkest parts while keeping the highest levels unaltered, as in Figure 8.4.

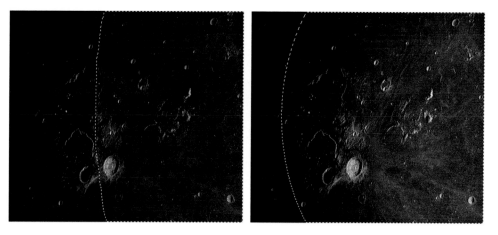

Figure 8.3 Left: selection with the lasso tool. The feathering/progressiveness of the selection is set to a large value, e.g. 20% of the image width. Right: during adjustment of the curves, the diffuse selection preserves the low-contrast area while the high-contrast area is brightened.

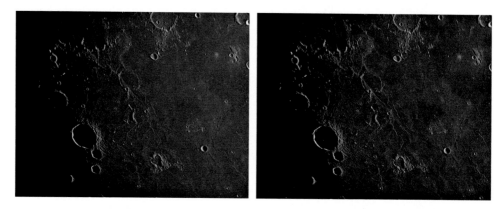

Figure 8.4 The contrast remains unchanged in areas with low dynamics (near the terminator) while it is expanded in the right part of the image. The maria now show conspicuous variations in albedo. On the other hand, the surface now seems planar rather than spherical (17° W, 18° S).

Diffuse selection was presented here for contrast enhancement. Of course, other processes may take advantage of it, for instance sharpness enhancement and noise reduction. The drawback is that differentiated processing of the parts of an image may lead to an unnatural rendering, with very sharp mountains and craters while dorsa and the terminator are excessively softened. Variations of diffuse selection are the "magic wand" and "color select": the selection is based on the levels rather than on the area.

8.3 Multiple-layer processing

Practice shows that contrast adjustment and sharpness adjustment may be incompatible. Actually, the adjustments are the same at different scales, and this doubles the noise. Moreover, some filters decrease large-scale contrast while the visibility of details is enhanced. This dilemma leads to an evident solution: the combination of two parallel processings of the same image. The image is duplicated and forms two layers. The first one is processed to emphasize large-scale contrast (maria, pyroclastics …) without affecting the details, that is, with little noise. The contrast may be enhanced by special functions, e.g. tone mapping, Fitswork[1] (Processing → Background flatten → Automatic flatten for nebula, then manual background flatten), GIMP (Retinex with "low" mode) or Photoshop® (Shadows/Highlights). The second layer is processed to reveal details like craterlets with no consideration at all for large-scale contrast. The best example is the particular sharpness filter of Fitswork (Processing → Special filter → Unsharp masking) with the "strength" parameter set

---

[1] This software is intended for processing deep-sky images (the documentation is in German).

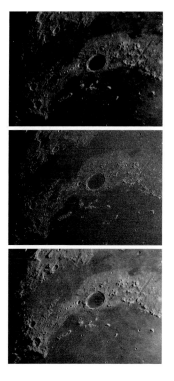

Figure 8.5 Top: a part of an image processed for large-scale contrast enhancement: maria and highlands are fairly differentiated. This reveals the dependence of the soil albedo on the illumination angle. Center: the same image independently processed with an unsharp mask for small-scale enhancement (mostly topographic features). Bottom: the combination of the two images as layers. Double-layer processing is a simple and elegant solution to get the best of both worlds with little effort (10° W, 48° N).

to 80% and the "radius" parameter set to two or three: large-scale contrasts are totally erased while small-scale contrasts are outlined. The layers are finally combined with a special mode that depends on the image editor ("multiply," "luminance," "hard light," "screen" . . .) (Figure 8.5). The goal is to use the large-scale contrast layer as a luminance multiplier of the second layer, which contains only small-scale contrasts. Two layers will suffice for practically all cases.

8.4 Unveiling ray systems

The purpose is to emphasize ray systems with contrast-gradient expansion rather than sharpness enhancement. Several techniques may be exploited, but the results are variable from one image to another: the best approach is to try them successively on the same image. Figures 8.6, 8.7, and 8.8 show a dual-layer processing with unsharp mask and wavelets.

- Tone mapping: this is intended to alter the rendition of levels to adapt the original image to printing or viewing. The dynamics may vary by a factor of several hundreds. Tone mapping is the application of a look-up table (a level-translation curve) to differentiate close levels. Compressing low and intermediate levels while the brightest levels are expanded generates a strong contrast that brightens mainly the ray systems.

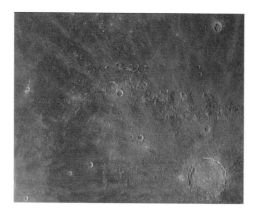

Figure 8.6 The image is copied as a layer, and then processed for sharpening of small-scale details only (27° W, 13° N).

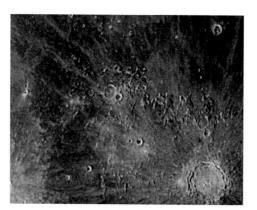

Figure 8.7 Levels 5 and 6 of wavelets emphasize only large-scale features. With RegiStax, depending on the image, dyadic wavelets may be more efficient. Small-scale details are unaffected, hence the noise remains low, but details are thickened, leading to a loss of accuracy. An appropriately chosen inflexion point stretches the differences in albedo.

Figure 8.8 The layers are fused. Various modes have to be tried to determine the best way to preserve the huge gradient of layer 2 while layer 1 restores the accuracy: "luminance," "multiply," "hard light," "darken" ... Other software packages provide equivalent combinations with arithmetic functions.

- Wavelets: the highest levels (levels 5 and 6 in RegiStax) help to increase large-scale contrasts only.
- Gradient map: this is a simplified version of tone mapping: a graphical, gradient chart is applied to alter the levels (e.g. Photoshop®: Image → Adjustments → Gradient Map, with an editable, monochrome gradient).
- Curve adjustment: this is the simplest way to compress low levels while expanding high levels. We have to find and stretch the inflexion point, that is, the intermediate levels where rays and the soil become clearly differentiated (this is easier with maria).

The original images have to be intrinsically well contrasted, preferably taken near the full Moon; prime-focus imaging guarantees the strongest possible contrast. In cases in which the terminator is relatively close, the process can be made easier by balancing the global gradient of the image beforehand.

8.5 Enhancing the contrast of pyroclastics

Pyroclastic deposits are areas covered by explosive ejections of very dark lava, and seventy-five of them have been listed. They represent unfamiliar and fascinating features because they testify to volcanic events and outpourings of magma. Since they are darker than maria, adjusting the level curves is sufficient to emphasize them, possibly with the help of the determination of an inflexion point (Figure 8.9). The lowest levels of the image should be reserved for pyroclastics, while the remaining regions of the surface, such as maria, highlands, and mountains, have to be leveled up in order for them to be clearly differentiated. The difference in albedo between maria and highlands does not need to be expanded because they do not constitute the main subject of the image. Some high levels must be kept in reserve if bright features – young craters, peaks – are enclosed in the image field, so that the global dynamic does not deteriorate excessively. Since they represent differences of soil composition, pyroclastics show no relief and very little differentiation in albedo when they are close to the terminator; therefore, there is no point in trying to unveil them under such conditions. A period close to full Moon is best.

8.6 Emphasizing wrinkle ridges, domes, and ghost craters

Here are the most elusive features of the Moon. With hardly any relief – altitudes barely reach 300 m or so (1000 ft)[2], gentle slopes – a few degrees – and absolutely no significant differences in albedo with respect to neighboring terrains (mostly maria), the only trick to unearth them is to wait for them to cross the terminator. Now we have to intensify a weak contrast between the shadows on the terminator side and brightness on the illuminated side. Of course, because the features show a very low dynamic, the contrast cannot be expanded without strongly increasing

---

[2] There are some exceptions, including Mons Rümker and Mons Gruithuisen gamma.

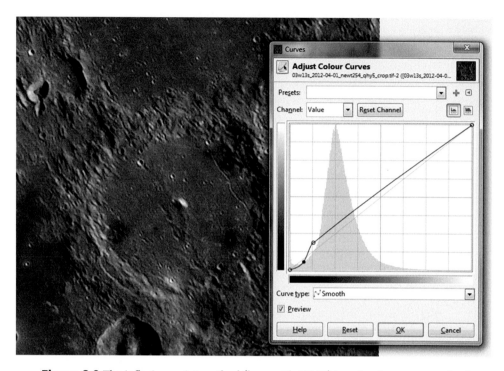

Figure 8.9 The inflexion-point method (here with GIMP) is a simple way to emphasize pyroclastics, as long as the region is illuminated at near right angles. The three famous pyroclastics in Alphonsus are related to rilles, indicating that they are likely centered around vents. The contrast adjustment differentiates them from their environment, leading to a relatively high level for the rest of the surface. Note that the pyroclastics remain lighter than the shadows. The result is shown in Figure 11.24 (3° W, 3° S).

noise and posterization. The solution is to stack numerous frames, for instance 1000, to enhance the signal/noise ratio. The impressive number of frames might seem astonishing, but this is the done thing in planetary imaging, where 4000 frames are commonly stacked!

These features are very smooth and little sharpness enhancement is needed. The golden rule is to maintain the darkest parts of the image just above absolute black, like in deep-sky imaging, and to raise the mare up to the highest acceptable level without introducing noise or posterization. The inflexion-point method is not recommended because there is little gradient; any noticeable contrast expansion would lead to posterization. Very smooth and effective gradient adjustments are provided by Photoshop®'s Shadow/Highlight or GIMP's Retinex. The purpose is to gently increase low levels while taking neighboring areas into account on a more or less large scale. What is especially interesting is a set of three parameters in Photoshop®'s "Shadows" box. The "Amount" parameter is the efficiency (e.g. 30%). The "Tonal width" and "Radius" parameters must be adusted in relation to each

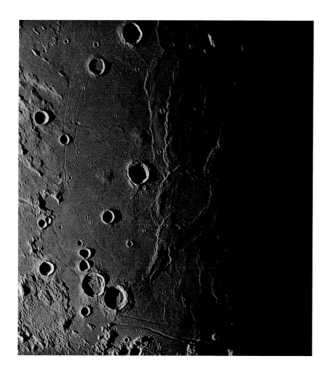

Figure 8.10 To the right of the ghost crater Lamont, an exaggerated increase in brightness produces dark lines, which are a side effect of the registration step while stacking. The brightness of the dark part of the Moon has to be lowered almost to zero. Even if they are hardly visible on a monitor, these dark lines may become evident on another monitor, or even more conspicuous if the image is printed in a magazine or with a home printer. The inflexion point is determined by the average level located exactly on the terminator; the other way is using functions like Shadows/Highlights (Photoshop®) or Retinex (GIMP). The image was taken with a noisy sensor (23° E, 5° N).

other. A low value (e.g. 20%) for "Tonal width" generally gives satisfying results. The value for "Radius" strongly depends on the image. The other parameters may remain at their default values. The drawback is the possibility of square vignetting, but this can be avoided by a preliminary enlargement of the canvas. The PixInsight software (https://pixinsight.com) has very efficient functions and scripts to process low-contrast images because it is intended for deep-sky imaging. If we do not have the required software, a diffuse selection may rescue the image by selecting a feathered band starting from the terminator before gently increasing low levels.

Working on such poorly reflective regions has a pitfall: raising up low levels may reveal some undesired traces of stacking in the form of black marks, outlining artificial, uneven tiles, depending on the stacking software (Figure 8.10). The lowest levels must be dimmed, at the price of a partial loss of the band of terrain located on the terminator. As for all images near the terminator, the adjustment in contrast is very tricky, and for this one requires a well-calibrated monitor.

8.7 Detecting and fixing rebounds

When we start to produce accurate images, there is a great temptation to excessively increase their sharpness with techniques such as unsharp mask, wavelets, and deconvolution. We have already seen how artifacts such as small-scale artificial patterns and tiles may happen. There is another important type of

Figure 8.11 Rebounds appear like echoes of contrasted features like the crests of young crater rims. They are caused by excessive unsharp masking/wavelets, especially in images taken with noticeable turbulence.

artifact, called rebounds. They resemble an echo at one or both sides of a bright and narrow feature, especially steep mountains and crater rims (Figure 8.11), and sometimes they even affect rilles and craterlets. Images taken with noticeable turbulence are more prone to rebounds because the alignment step during stacking cannot be carried out perfectly. This is a classic artifact when sharpness processing is very strong, to the extent that some software packages (e.g. RegiStax and PixInsight) include a specific corrective function called "de-ringing," which is primarily intended for planetary imaging (rebounds appears at the limbs of planets and at the edge of Saturn's rings).

The corrective action is tedious and arbitrary because it consists of manually erasing rebounds. As we become more experienced, it becomes easy to detect such rebounds. For circular features like young craters, both detection and correction are simple. Correction is feasible with a "clone stamp," "burn," or "clone brush" tool, depending on the image editor, or by a diffuse selection of the rebound prior to darkening it to zero. Manual fixing is problematic with non-circular features, such as ruined craters or irregular reliefs. If we are aiming for aesthetically pleasing pictures, manual fixing of rebounds is annoying and imprecise but acceptable, because this leads to very clean images. If we are trying to get precise images in order to study lunar topography, any interpretation must be considered non-scientific. Even the most talented astrophotographers may have small rebounds due to processing, in particular in planetary photography (e.g. at one limb of Mars or Venus, or around Jupiter's clouds). Actually, collapsed terraces in internal crater rims, like in Copernicus, may be confused with rebounds, along with hills on highlands. Small craters also seem to be concentric (real concentric craters do exist, such as Hesodius B, but the internal ring is visible at any illumination angle and whatever the processing) or to have a peak (real central peaks appear only in large craters, those of diameter at least 70 km (43 miles)). All these likely sources of confusion mean that one should avoid manual fixing, in favor of a moderate sharpening process.

8.8 The colored Moon

Recommended imaging device	Recommended optics	Shooting difficulty	Processing difficulty

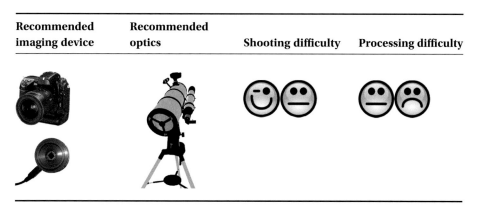

A color camera can reveal the very subtle tints of the soil. The most colored areas are the Aristarchus Plateau, Tycho, the boundary between Mare Tranquillitatis and Mare Serenitatis, and Oceanus Procellarum. The simplest way to reveal these colors is by using a DSLR (Figure 8.12) or a planetary camera with a large color sensor (e.g. ICX178) with a telelens, a small apochromatic refractor, or a small reflector. A focal length of 1200 mm is perfect to frame the entire Moon with an APS-C sensor. The basic process consists of a strong increase in saturation.

Paradoxically, a monochrome camera is a better choice, thanks to its versatility when it is used in combination with filters, as in deep-sky color imaging, especially because it can partially fix chromatism by choosing appropriate color bands and re-focusing the blue image. Here are several possibilities.

- RGB: three monochrome images are successively taken with a red, a green, and a blue filter. The images are merged into a color image (Figure 8.13).
- RRGB: the monochrome image taken with a red filter is less altered by the turbulence. The three other monochrome images with red, green, and blue filters are merged into a color image, then the latter is superimposed as a chrominance layer onto the luminance image. This is the best solution when the turbulence is noticeable (Figure 8.14).
- LRGB: this is the same technique as above but with the luminance (L) layer, which is often taken with a luminance filter (that is, a clear filter with infrared

Figure 8.12 Left: the original image taken with a small reflector and a DSLR. Right: a simple – but unmoderated – increase in saturation reveals the colors.

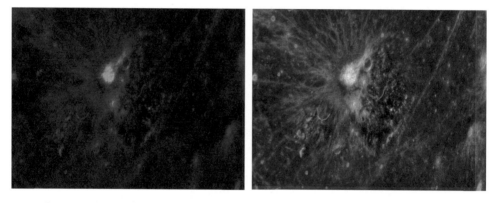

Figure 8.13 Left: the three monochromatic layers taken respectively with red, green, and blue filters and a monochrome camera are colorized, and then roughly super-imposed with an image editor. Right: the final steps are performed with RegiStax: RGB align, RGB balance, and then wavelets (49° W, 23° N).

and ultraviolet rejection). If the Moon is low, an ADC substantially improves the quality of the luminance image. This solution is best for taking short exposures (especially in the case of an unmotorized mount).

To ensure that the colored areas are valid, the images can be compared with professional images. Figure 8.14 shows the boundary between Mare Serenitatis (North) and Mare Tranquillitatis (South). The first one contains iron oxide (FeO), whereas the second one contains FeO, ilmenite ($FeTiO_3$), and titanium dioxide

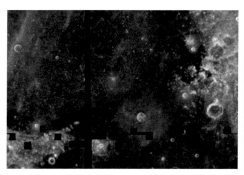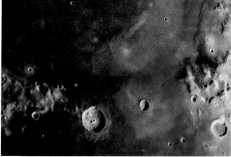

Figure 8.14 Left: An ultraviolet-plus-visible image taken by the LRO space probe; black areas are missing data. Right: an RRGB image (red as luminance plus colors with RGB filters) with a monochrome camera and a 10-in Newtonian in poor, turbulent conditions. The processing in "natural" colors is serendipitously consistent with the LRO image (25° E, 17° N). Left image by NASA/GSFC/Arizona State University.

(TiO_2), and the soil is older. These differences in age and composition, in addition to various events such as micrometeorite impacts, effects of the solar wind, cosmic rays, and pyroclastic flows (such as Taurus–Littrow to the East), alter both colors and albedo.

Another means to emphasize the differences in colors is arithmetical processing (difference, logarithm, multiplication … are functionalities provided by image editors). This enhances the readability of the image, but it takes a long time to determine which processes are the most interesting. The intended result may be unnatural, but false colors improve readability.

Some attempts have been made with narrow-band imaging, that is, using specific, very-narrow-bandwith filters primarily intended for imaging nebulae (such as O III, S II, and H II filters). The idea is to more easily separate the colors. The results were somewhat disappointing, very similar to those obtained with classic RGB filters, because most of the differentiation manifests itself in the near infrared, especially between 930 and 1100 nm, where the red-side depletion of most astronomical and industrial cameras occurs.[3] Another problem is that such narrow-band filters eliminate an enormous part of the light (this is what they are supposed to do) and dramatically dim the image, leading to long exposures – in terms of lunar imaging. A more professional way to image the Moon in color is petrographic imaging with specialized filters. This requires you to "subtract" the maturation ratio to eliminate a bias from the differentiation of terrains by age, because the solar wind alters the composition of the surface with time. Neglecting the maturation ratio could lead to a confusion between the age and the composition.

---

[3] Performance is improving with sensors such as E2V's near-infrared-depleted Ruby and Sony's IMX224.

Figure 8.15 A stack of two rotating polarizing filters is intended to serve as an adjustable neutral filter to dim the too bright full Moon. It can reveal the faint variation of the global polarization of the solar light reflected by the lunar surface in correlation with the phase.

Spectacular results have been reported by experienced amateurs: see www.astrosurf.com/lecleire/mai2007.html by J.-M. Lecleire (France) in 2007 with a very interesting comparison in the red, green, blue, and 685-nm and 807-nm near infrared (in French); and, more recently, www.rkblog.rk.edu.pl/w/p/my-filters-and-first-results-lunar-petrographic-imaging by Piotr Maliński (Poland), including mafic (magnesium plus iron) and maturation images.

A faint variation has been observed in polarized light, which depends on the nature of the surface with respect to the phase.[4] This has been useful as a means by which to compare the lunar soil with terrestrial basalts independently from the first robotic and human landing missions to the Moon. The experiment is feasible with an amateur polarizing filter (Figure 8.15), intended to dim the dazzling full Moon when it is observed with a 150-mm (6-in) telescope or larger and a low

---

[4] E. Bowell, A. Dollfus, and J. Geake, 1972; for further reading, see www.lpi.usra.edu/meetings/lpsc1990/pdf/1154.pdf.

magnification. It consists of a stack of two polarizing filters; when we rotate the first one relative to the second, less light passes through. A polarized beam of light is in a certain sense "orientated" at right angles to its direction of travel. When the orientation matches the orientation of the polarizing filter, the light passes through the filter.[5]

The results were disappointing (for amateur imaging), because the variation is not pronounced; the main effect is that the polarization varies with the phase, and the variations are too strongly correlated to provide a new view of the Moon. But this approach may merit further investigation.

8.9   Inverting the image

Inverting the image is a legacy of deep-sky imaging. Tenuous extensions of galaxies and nebulae can be more visible on a negative image, because our eye more easily detects dark details on a bright background. Inverting or excessively expanding the contrast and colors of the Moon helps reveal interesting features such as the soil composition and age or ray systems. This is useful to draw attention to otherwise barely detectable features. A good example is a "psychedelic" saturation increase (Figure 8.16) to reveal the boundaries of the differentiated areas of Mare Serenitatis and Mare Tranquillitatis. Another good

Figure 8.16 Inverting the image emphasizes ray systems and differences in color and albedo, because our eye more easily distinguishes tenuous, dark features on a white background. The same process with an exaggerated saturation of a color image emphasizes the differences in age and composition of the surface.

---

[5] The same technique is exploited in 3D movies: our left eye receives the polarized light in one orientation while our right eye receives the polarized light at right angles.

application is preparing global charts of maria and highlands by exaggerating the differences in albedo and outlining their boundaries.

8.10 Making a giant image: the mosaic

Recommended imaging device	Recommended optics	Shooting difficulty	Processing difficulty
		☺	☺

When we wish to obtain a very large image, we can either use a large sensor (a DSLR or megapixel camera) or create a mosaic. Within three minutes, the colongitude remains practically steady and several movies of the same area can be acquired without noticeable variations in lighting. The movies must have common edges, so that the final images can be joined. Since the acquisition parameters (exposure, gain, orientation of the camera …) have to be exactly the same, they should be set according to the brightest area, avoiding overexposure and clipping, even if the darkest areas are underexposed. The processing must be rigorously identical for the different areas. Some acquisition software (e.g. FireCapture) helps to frame the successive areas while keeping common edges. Otherwise, young craters provide convenient markers. If gusts or irregularities in tracking are foreseen, a margin of 20% in width is a good precaution.

Classic image-editing software packages are able to superimpose the images as semi-transparent layers, but in practice the edges are distorted by the registration and stacking step, not to mention possible field curvature, field distortion, and peripheral degradation (the edges are computed with fewer frames unless the tracking was absolutely perfect). A correct alignment demands a morphing capacity. This is how software packages dedicated to panoramic photography work. Some are provided with consumer DSLR cameras (e.g. Photostich with Canon). Others are free, e.g. Microsoft ICE, Hugin, and iMerge (Appendix 2), which was specifically developed for lunar mosaics (Figures 8.17, 8.18, and 8.19) by Jon Grove. iMerge loads images up to 1024 pixels in width; the formats are limited but convenient for the Moon: BMP, FITS (32 bits), even AVI.

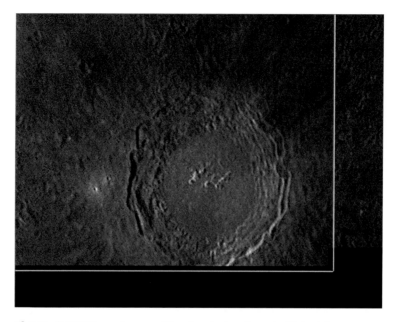

Figure 8.17 iMerge is one of the simplest and most efficient freeware programs to create mosaics, thanks to its real-time morphing and lightness homogenization. The limit of each individual image is 1k pixels in width. The assembled result may be very large, e.g. 4k or 8k, which is enough for a large print in an exhibition.

8.11 HDR images with software or manual combination

Recommended imaging device	Recommended optics	Shooting difficulty	Processing difficulty

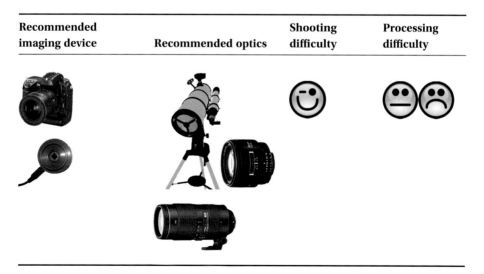

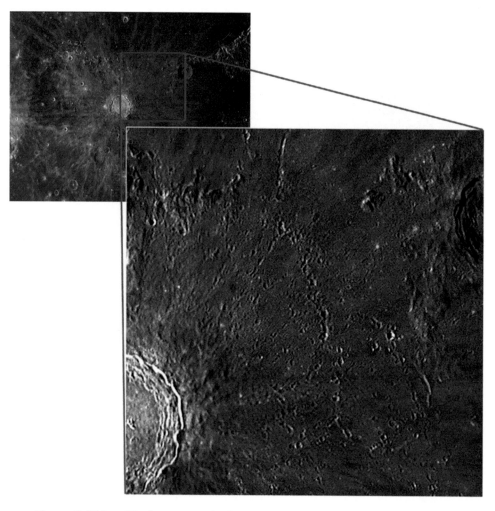

Figure 8.18 Detail in the mosaic. The format is 2195 × 1674 from six frames acquired with a 1024 × 960-pixel planetary camera.

Few sensors are able to manage high dynamics (a very wide range of light levels without overexposure or underexposure). Another technique consists of taking different exposures within a small time interval and then combining them as a single image, containing both highest and lowest levels. A handy way to combine different exposures is by using software that includes a high-dynamic-range (HDR) feature, for instance Photoshop® CS5 or a later version, Photomatix, Fusion, AutoHDR, PhotoHDR, easyHDR, LuminanceHDR (formerly "qtpfsgui"), Akvis HDRFactory, PTGui, Hydra, EnfuseGUI, Photoroom, Picturenaut, PFSTools, FDRTools … or specialized, astronomical imaging software, e.g. PixInsight. Some of these software packages are available in a try-and-buy version with

Figure 8.19 The full Moon shoot at the F/D = 8 prime focus of a 150-mm (6-in) apochromatic refractor with a 1/2-in sensor. The mosaic contains nine images, resulting in an HD, 2k image.

a watermark; some are freeware. The results are often dramatic and easy to obtain. Nonetheless, we have to deal with three problems.

- Some software cannot align images well (there are vertical, horizontal, and rotational shifts between exposures, especially with no tripod or with an unmotorized mount). Prior to processing the different exposures with the HDR software, it is better to manually align the images with an image editor.
- HDR software often uses EXIF data, in particular exposure times. EXIF data are embedded in photographs, either JPEG or RAW. When we align images

beforehand, the original EXIF data must be preserved. For instance, if we align the first image with respect to the second image, we have to superimpose the second image as a semi-transparent layer, then align the background (the first image transformed into a layer), then delete the second image layer, and finally save the first, re-aligned image, keeping its original EXIF data.

A classic bane in HDR software is "ghost effects": artifacts such as halos, disks, or rings may appear around the brightest areas. HDR software often provides built-in functions to wipe out such artifacts, but some of them may remain, especially light or dark circles around the Moon. That is why, despite astonishing results with daylight images, such software may turn out to be poorly suited to lunar photography. In fact, the difference in brightness between the full Moon and the background of the sky (a light-polluted urban sky, or a dark sky lit by the full Moon) is about 16 magnitudes, that is, a factor of 2.512^{16}, or 2.5 million![6] This requires, at least, 22-bit images (for each color), far more than our 12- or 14-bit DSLR raw photographs, not to mention 8-bit JPG photographs. Even deep-sky cameras have no more than a 16-bit capability, that is, 65 536 levels, almost 40 times less than a 16-magnitude difference. HDR software packages are not intended to manage such huge lightness gradients, and this is why sensors should be able to provide built-in HDR capabilities. For instance, the Aptina MT9M034 color sensor has a 20-bit HDR capability with an automated, two-pass exposure. This capability of 20 bits means that the sensor could almost entirely accept the lightness gradient between the background level of the illuminated sky and the full Moon. Unfortunately, this capability is not operational at the time of writing, with some exceptions, such as QHYCCD's QHY5 L-II with a 20-bit to 12-bit companding function. Other sensors require an additional image signal processor, possibly developed after the cameras had become available. Indeed, because of artifacts from HDR software, and despite the author's fondness for HDR images, he generally uses classic image editors with "manual" superimposition of multiple-exposure photographs. Nonetheless, if images show limited dynamics, HDR software performs well and provides dramatic final images.

8.12 Tone mapping

After we have obtained an HDR image, we are faced with a classic problem: the final medium, that is, a computer screen, photographic quality paper, or an offset-printed issue of our favorite astronomy magazine, has a very limited range of distinguishable levels. The worst case is when we want to print an HDR image that

[6] The magnitude scale was empirically determined by the ancient observer Hipparchus from 1 (brightest stars) to 6 (faintest stars). With modern measurements, this scale was more accurately redefined, even expanded to negative magnitudes (the Moon's magnitude is –12.6). The difference between two magnitudes is a factor of 2.512 in terms of light flux.

contains, for example, a million levels, on ordinary paper, which can reproduce no more than a hundred differentiated levels.

The solution is to reduce the range of levels in order to adapt it to the final medium. The simplest idea is to apply a linear conversion, such as conversion of a 16-bit image (65 536 levels) to an 8-bit image (256 levels) with the help of an image editor. With a true 32-bit HDR image, this implies the reduction of 4 294 967 296 levels down to only 256 levels! Of course, a significant loss in subtle shading is to be expected. This is especially noticeable in low-light areas such as maria and ridges near the terminator. This is why a simplistic, linear reduction of the range of levels is not appropriate. As an alternative, a special rule for reducing the range can be adopted, in the form of a conversion table, or look-up table (LUT). The whole range of levels of the original image is broken down into non-linear intervals of levels. Similar values can be merged into a single, mean value. But we may want to preserve subtle differences between similar values in some intervals. By adopting uneven intervals, determined empirically and depending on the shading we want to keep (especially the highest values for crater rims and lowest values for maria), we can adapt the range reduction to our needs. For modifying the conversion rules from a wide range to a small number of significant, differentiated intervals of light levels, we utilize an LUT. This technique is tone mapping (Figures 8.20 and 8.21).

Tone mapping may be a side effect with some software, by simply reducing the color depth of an image, for instance to convert large images with numerous colors into smaller images (in terms of file size/kilobytes) with only 256 colors for the Web. Linear reduction of colors and diffusion to simulate nearby colors are often used. More sophisticated image editors can do much more. For instance, if we load a 32-bit image in Photoshop® and then convert it to 16 bits (menu image → mode → 16 bits/channel), a dialog window asks how to convert the range of levels

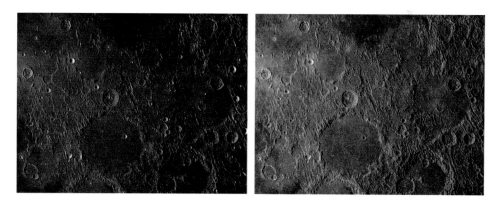

Figure 8.20 Left: the original image. Right: tone mapping (here with LuminanceHDR) expands subtle differences (1° W, 7° S).

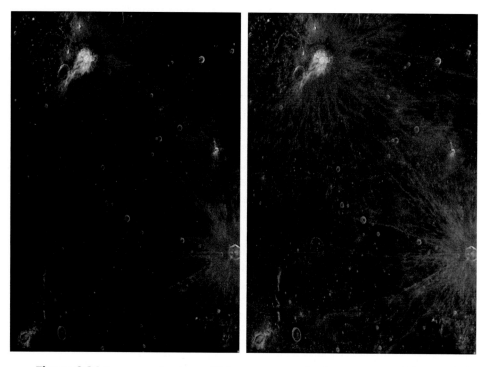

Figure 8.21 Tone mapping is an efficient way to emphasize ray systems (along with the Reiner gamma magcon). Left: original image. Right: tone mapping (50° W, 15° N).

of the HDR image into an image with a smaller color depth. If the original image is already in 16 bits, use menu image → mode → 32 bits/channel prior to reducing it to 16 bits. The most interesting method is selecting "local adaptation" and opening the "toning curve and histogram" tool. Modifying the curve, by creating any number of inflexion points (by clicking anywhere along the curve) and then moving these points horizontally and vertically, allows manual design of the LUT. The curve can be saved and reused to apply the same dynamical reduction to other images. After the reduction, the final, 16-bit image can be saved with a correct adaptation of levels. Of course, the more specialized software mentioned in Section 8.11 includes a tone-mapping function. As another example, Photomatix Pro provides series of presets for tone mapping (menu File → open, then open the image, then click on "tone map . . . "). Pfstmo (Linux) is another free, complete package for tone mapping, provided with the qpfstmo graphical interface (the spellings are correct!). LuminanceHDR is also very interesting in many ways: it is free (a good opportunity to get acquainted with using HDR and tone mapping) and multi-platform (OS X, Windows, and Linux); and it supports various image formats, including the astronomical FITS image format.

9

Making 3D lunar images

9.1 Three-dimensional anaglyph

A 3D anaglyph is a stereoscopic image, formed of images taken from two slightly shifted viewpoints. The differences in perspective between the two images are encoded and represented in cyan and red, respectively. To recreate the feeling of stereoscopy, one has to use cyan and red eyeglasses. Basically, the most obvious way to make a lunar anaglyph is to image the Moon from two sites several thousand kilometers apart at the same time. This has been done successfully by a few amateurs. Another way is to exploit the librations. It is normally not possible to wait for the Moon to rotate between two shots, because of the continuous evolution of its phase, bringing unwelcome variations in illumination of the topography, but the phase is less pronounced near the poles, as we can see at

http://www2.lpod.org/wiki/August_11,_2008

Three-dimensional anaglyph of the Moon

Yuri Goryachko, Mikhail Abgarian, and Konstantin Morozov (Belarus) imaged the region containing the Bailly and Hausen craters twice in 2008 (June 27 and July 25) with a 230-mm Santel Mak, a near-infrared filter, and a Unibrain Fire-i 702 camera (Figure 9.1). They assembled the mosaics to create a stunning 3D anaglyph with the help of librations (Figure 9.2).

   "Astronomy has been our favorite hobby since childhood," Mikhail explains. "Yuri and Konstantin began film astrophotography prior to forming our group in 2007. Our main instrument, a Santel, 230-mm Maksutov–Cassegrain, was replaced in 2013 by a 360-mm Klevtsov–Cassegrain from the same manufacturer; only a single instrument was built. The Moon has always been one of the most interesting targets for us – this seemingly unchanging and boring object actually gives ample space for experiments. In addition to 'standard' shooting of lunar landscapes, our particular passions are full Moon mosaics and color images. This material collected for a long time allows us to obtain stereo images. For now our main goal is to realize mosaics for all phases at prime focus of the new telescope. Our dream is to create a mosaic of the full Moon with a focal length of 8–10 meters."

Figure 9.1 The Astronominsk imaging group. From left to right: Konstantin, Yuri, and Mikhail.

Figure 9.2 This dramatic image is viewable with red–cyan glasses. The image has been cropped to fit the format of the book. The best area was preserved (52° W, 66° S).

Mikhail, Yuri, and Konstantin's images have been published several times in LPOD and Sky & Telescope. They have a web site at www.astronominsk.org/index_en.htm.

A selection of software for anaglyph creation is available at

www.stereoscopy.com/downloads

Another method involves casting a single image onto a virtual sphere in special 3D software (e.g. Blender or 3DSMax®) and then slightly rotating the sphere. In this way, the two shifted images can be processed by anaglyph-generation software such as Anaglyph Maker:

www.stereoeye.jp/software/index_e.html

Figure 9.3 was realized by rotating a full Moon image with the 3D-transform function in Photoshop®.

Figure 9.3 This stereoscopic image results from a single image virtually cast twice onto a sphere with a slight rotation. The two shifted images are merged into a stereoscopic anaglyph with the help of Anaglyph Maker.

In place of red–cyan glasses, the author had already tried 3D video eyeglasses that had been designed primarily for simulators; the resolution was not accurate enough for still, high-resolution images.

9.2   Spherical projection with Photoshop®

A 3D "filter" in Photoshop® transforms any image into basic, geometric shapes. One of them is a sphere, a convenient basis for the Moon. Once we have loaded our image of the full Moon, or any other phase, we can transform it into a 3D object, and then rotate it around its central axis to see the Moon at very unusual angles (Figure 9.4). Starting from the main menu, we click on

$$Filter \rightarrow Render \rightarrow 3D\ Transform$$

We need only select the "Sphere Tool" (we can hover over the icons to display a legend about each icon), drag from top left to bottom right to entirely encircle the

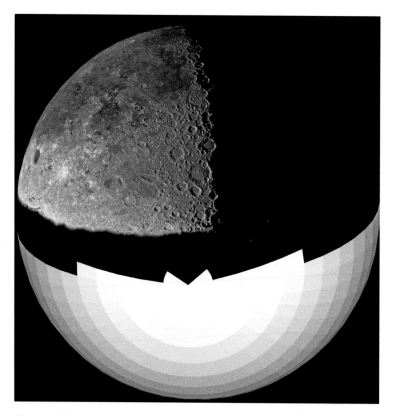

Figure 9.4 The 3D Transform filter of Adobe Photoshop® easily transforms an image into a sphere we can rotate. Here is the Clavius crater, virtually viewed from the lunar South pole. The original image is at the last quarter.

full Moon, click on the "Trackball Tool", and then, when the mouse cursor looks like a curved, bidirectional arrow, drag with the mouse in any direction to rotate the Moon.

Not only are such 3D transformations nice to look at because they are unusual – we can now see Clavius or Plato from above – but also they provide valuable information about the real shapes of particular craters such as Messier and Schiller, the orientation of Vallis Rheita, or elusive, tangent features like Mare Crisium, Mare Marginis, and Mare Smythii.

9.3 Albedo-based 3D

This technique is rarely used because it does not preserve the topography, although it may provide interesting perspective views of regions that we normally see only at right angles. The basic idea is to consider lighter areas as higher areas. This is true for bright highlands and dark maria on a broad scale, but not for such details as craters: shadows on the dark side of the rims and floor will be interpreted as depressions, while illuminated parts will be interpreted as peaks or rims, resulting in a slightly chaotic relief (Figure 9.5). At full Moon, flat features such as ray systems will appear as bumps, while small, fresh craters will look extremely

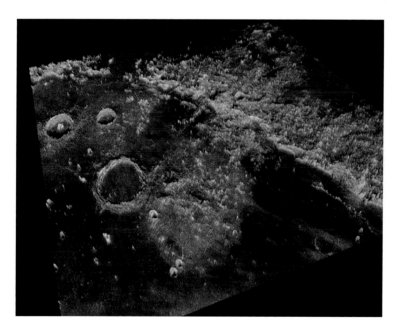

Figure 9.5 An albedo-based relief image. The Fitswork software by Jens Dierks is able to transform a photograph into a perspective view with relief, interpreting lighter areas as higher areas. This function is accessible via the menu: Processing → More functions → 3D Pixel view. A 3D software package such as Blender offers comparable features (3° W, 24° S).

245

abrupt. Craters imaged during a phase, with an internal and an external wall lit up while the other walls are in darkness, will appear with the shape of a spoon.

Some astronomical software is able to build an elevation image, for instance to show a spiral galaxy in relief. The author finds this function not particularly useful; it can thoroughly rearrange lunar images in perspective and highlight steep slopes.

9.4 Casting a photograph onto LOLA's DEM

This technique is based on virtually casting a real photograph onto a digital elevation model (DEM), that is, a digital representation of the relief (Figure 9.6). A DEM is a grid where numbers represent the altitude at each point. It can be represented as an image, where bright areas are high-altitude mountains and crater rims, while dark areas are crater floors, maria, and rimae. An accurate model has been established by NASA's Lunar Reconnaissance Orbiter probe and its on-board laser altimeter. We can find the DEM at USGS:

www.mapaplanet.org/explorer/moon.html

It's easier to choose a region of which we have a telescopic straight-on view in order to minimize distortion by the viewing perspective. Near the limb or the pole, it would be necessary to distort the photograph as a part of a sphere, before casting it on the DEM.

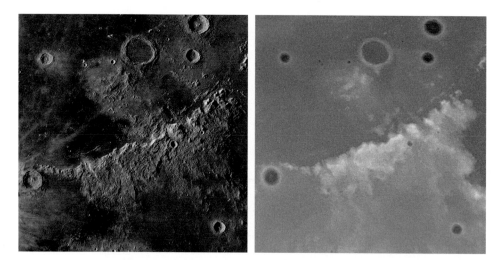

Figure 9.6 The two necessary components to construct a realistic 3D image: a real photograph and a DEM. Left: the photograph reveals strong differences in albedo due to soil composition, illumination angle, and topographic features. The latter cannot be inferred solely from the albedo, so for reconstructing the relief one needs an additional, reliable topographic model. Right: a view of the same region as a DEM from the Lunar Reconnaissance Observer Laser Altimeter, by LOLA Science Team, Goddard Space Flight Center.

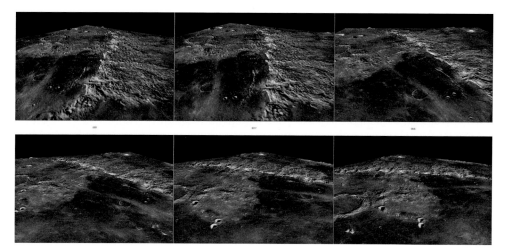

Figure 9.7 Frames from a computer-generated movie with the previous DEM from LOLA Science Team and photograph. Altitudes are represented with little amplification in order to produce a realistic image (7° W, 27° N). 3D processing with Blender by Mathias Barbarroux.

The author tried to use the renowned Blender freeware for 3D image generation. By casting the photograph onto a "bump map" (the relief map, which was here a crop from the global LOLA DEM), the image was correctly, albeit not very dramatically, rendered, because a bump map is intended to reproduce the roughness of a surface rather than to represent topographic features. Then the author asked a professional infographist to realize a short movie (Figure 9.7) from the photograph and the DEM. The DEM was transformed into a 3D grid. A distortion was applied to the photograph beforehand to compensate for the deformations caused by the librations and the angle of view from the observing site. This was done by visualizing the DEM and the photo as semi-transparent layers, in order to apply some orientation, stretching, and resizing for the purpose of matching features of the photograph to the DEM. Peripheral features were not aligned perfectly though.

This powerful technique allows us to appreciate lunar regions from new points of view. Since amateur images are not very accurate (e.g. 0.7-km-wide details), low-resolution crops of the global DEM from LOLA are perfectly adapted to our needs. Note that shaded DEMs, comprising color-coded points for altitude and shadow effects to enhance human-eye appreciation of the topography, are not applicable, because the shadows alter the interpretation of altitudes. Taking two shots, with a slight shift in viewing angle, allows the derivation of 3D anaglyphs. Creating such images requires a good knowledge of 3D imaging, but astronomy clubs may be a good place to learn this exciting technique, or to establish a collaboration with infographists with the help of forums.

9.5 Creating a modern Wright globe

Representing a 3D Moon from photographs was a challenge for manual techniques, because photographs are planar while the real Moon is spherical. In the 1930s, Dr. Frederick Eugene Wright projected lunar negatives (plates) taken with the 100-in telescope at Mt. Wilson Observatory onto a globe coated with photographic emulsion to produce a spherical positive. This was the only way to see regions near the poles and the limb, which are normally narrowed by perspective. We can see a so-called Wright globe here:

http://lpod.wikispaces.com/August+17,+2013

Modern techniques allow us to easily create a real, or virtual, Wright globe, and see peripheral regions as if we were flying over them.

9.5.1 With a videoprojector

This is a fun way to have a giant globe of the Moon. We always start with a photograph of the full Moon. A video projector casts the image on a "gym ball," "exercise ball," or "fitness ball": an inflatable ball with a diameter of 55–75 cm (22–30 in); sport sometimes helps science (Figure 9.8). Despite the surface not being a projection screen, provided that the gym ball is not too dark and has a single color, the photograph can be projected onto the sphere.

An original Wright globe, being coated with photographic emulsion, recorded the image. This is not the case with a gym ball. The author played with an old technique called gum bichromate, requiring dangerous chemicals and an ultraviolet-light bulb in the photographic enlarger. The main problems are that the

Figure 9.8 A videoprojector casts a photograph onto a green, 60-cm gym ball. Right: we can see Mare Crisium from above, a view impossible from Earth. The gym ball was attached to the ceiling by a wire wrapped around the valve. The inflatable ball is slightly egg-shaped because it is stretched downward by its own weight.

sensitive coating reacts to light according to the thickness of the coating, and the accuracy is not very good. The main advantage is that the coating (made with gum arabic) may be applied to almost any resistant surface, not just paper. The best way to immortalize our Wright globe is to shoot it at various angles, especially from above near limbs and poles.

9.5.2 With WinJupOs

WinJupOs is designed and freely provided by Grischa Hahn at

www.grischa-hahn.homepage.t-online.de/index.htm

Devoted to planetary photography, one of its most striking features is the ability to internally cast each movie frame to its original location on a virtual, rotating globe. In other words, this software can cancel the blurring effect of the rotation of a quickly rotating planet such as Jupiter. WinJupOs can also reconstruct polar and cylindrical maps of the planet from a mosaic of photographs (Figure 9.9). This means that, with a single image of the full Moon, we can create a virtual Wright globe.

First, we need to transform an image into a cylindrical map, which is a standard representation for planet mapping.

- First, click on the WinJupOs menu Program → Celestial body → Moon.
- Then on recording → image measurement.

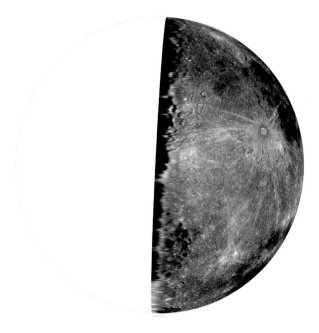

Figure 9.9 Polar projection of our image of the full Moon with the help of the brilliant Grischa Hahn's WinJupOs – an impossible view from Earth.

- In the "Imag." tab, select "Open image" and then load the image of the Moon.
- In the "Adj." tab, match the circle with the shape of the Moon, using the following keys:
  - (kbd-arrows.tif) [←][↓][→] for movement;
  - (kbd-PgUpPgDn.tif) [PgUp]/[PgDn] page Up/Down for size;
  - the keys may be hit in combination with (kbd-shift.tif) [⇧] or (kbd-ctrl.tif) [Ctrl].
- Back in "Imag." tab, enter the date, time, and coordinates of the observation site from which the photo has been taken, then click on "Save".

Then we can ask the software to compute a polar or equatorial point of view including the whole sphere (bar the far side).

- Click on Analysis → Map computation, then apply the following parameters:
  - Projection type = Polar projection
  - Latitude scale = Planetocentric
  - Map orientation = South pole at top (this depends on the orientation of our image)
  - Layout/Longitude = −180°
  - Layout/Automatic brightness . . . checked
- Then click on "Compile map" button.

9.5.3 With Celestia

Celestia is a renowned freeware package to simulate the Solar System, stars, and galaxies. It can load third-party, cylindrical maps and cast them onto planets and other celestial bodies. This provides a convenient way to reconstruct the Moon from an amateur image. Firstly, we have to ask WinJupOs to reconstruct a cylindrical map from our image of the full Moon (Figure 9.10).

- Click on Analysis → Map computation, then apply the following parameters:
  - Projection type = Cylindrical projection
  - Latitude scale = Planetocentric
  - Map orientation = North pole at top (this depends on the orientation of our image)
  - Layout/Longitude = 180° and Automatic brightness . . . checked
  - Then click on the "Compile map" button.

Once the map file has been created, we have to import it into Celestia. The cylindrical maps are in the "Celestia\textures\hires" folder. The "Celestia" folder may be located in various places, depending on the operating system (e.g. "C:\Program files (x86)\") (Figure 9.11).

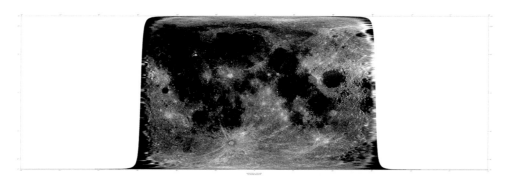

Figure 9.10 A cylindrical map made with WinJupOs, from our image of the full Moon. This step is necessary so that we will be able to ask Celestia to cast the map onto a virtual sphere afterwards.

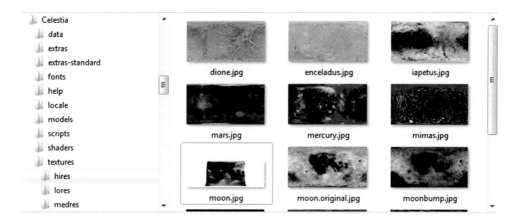

Figure 9.11 In the Celestia texture\hires folder, rename the original cylindrical map "moon.jpg" as "moon.original.jpg," and copy the cylindrical map made with WinJupOs to "moon.jpg." To restore the initial behavior of Celestia, delete "moon.jpg," rename "moon.original.jpg" as "moon.jpg," and then restart Celestia.

(1) Open the "Celestia\textures\hires" folder, then rename "moon.jpg" as moon. original.jpg."

(2) Copy a jpg version of the cylindrical map into Celestia\textures\hires as "moon.jpg."

(3) Run Celestia, go to the Moon, look at it, and then accelerate time until the non-blank part of our map is lit up by the Sun (Figure 9.12).

Figure 9.12 A real-time, animated spherical projection of the Moon with Celestia. This requires the cylindrical map of the Moon calculated by WinJupOs from our image of the full Moon.

9.5.4 Spherical rectification

This is not really a 3D visualization tool, but this processing allows a virtual view, at any angle, of any location of the near side. The idea is, once again, to virtually cast an image onto a virtual sphere. The purpose is to correct for the perspective. Thus, the famous Plato crater, for instance, will no longer appear elongated. Almost any 3D software can do this, but there exists a specific tool to accurately visualize lunar features from above:

http://ltvt.wikispaces.com/LTVT

This is a scientific tool, freely provided and exhaustively documented. Furthermore, it is able to take into account a DEM to calculate the rectified view.

10

Measuring and identifying lunar features

10.1 Horizontal resolution

Measuring the diameter of craters is the most reliable means to estimate the accuracy of images, but we have to choose the right crater. Since large wandering bodies of the Solar System became rare 3.5 billion years ago, few craters have formed recently. The size of ruined craters is hard to estimate. Fortunately, numerous small craters are young and they are an appropriate choice, especially young, bowl craters, with sharp edges. The selenographic position alters the shape of craters: if the latitude is close to the pole or the longitude is close to the limb, the crater is narrowed vertically and/or horizontally (Figure 10.1).

The Virtual Moon Atlas (Appendix 2) offers an extremely handy measurement tool, taking into account the librations and the sphericity: it performs a geodesic measurement. By comparing the actual size of bowl craters in the image and in the software, we can estimate the resolution. The angular size of the crater is calculated by the software, assuming that the date has been provided correctly in order to compute the ephemeris. Another powerful tool is NASA's Act-React QMAP (Appendix 2), as shown in Figure 10.2.

10.2 Vertical resolution

Ever since Galileo Galilei, cast shadows have been exploited to attempt to measure the heights of lunar features. The calculations are a bit complicated, and the result is not very accurate because measuring shadows on the Moon is not easy. But nowadays we can use the marvelous tool mentioned above, NASA's Act-React Quick Map – abbreviated as QMAP – based on altimetric maps established with the help of the LOLA laser altimeter aboard the Lunar Reconnaissance Orbiter (LRO) space probe. The only thing we need to do is draw a line on the map to obtain the altimetric curve (Figures 10.3 and 10.4).

Such altimetric measurements avoid misinterpretations. The age and the composition of the soil are quite different at this location (Section 8.8). Thus, the darkest area is not solely due to a shadow of a gentle slope.

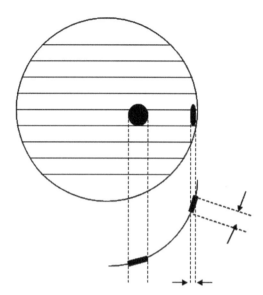

Figure 10.1 Perspective distorts the shape of the craters. If the feature is located near the latitude 0°, just the horizontal dimension is altered while the vertical one remains reliable for measuring the crater.

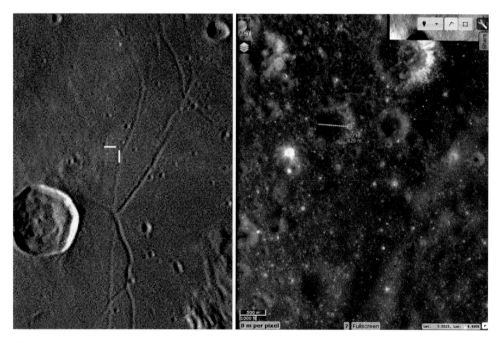

Figure 10.2 Measuring the resolution of an image (left) by comparison with QMAP (right). The smallest crater (white, half cross) is here 0.6 km wide. This is consistent with the theoretical observing resolution of 1.0 km. The crater cannot be confused with noise because it is round and shows a lit side and a dark side (4.5° E, 5° N).

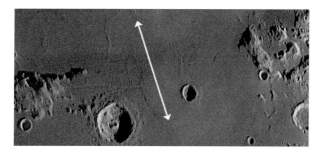

Figure 10.3 Choosing a slope at the boundary between Mare Serenitatis and Mare Tranquillitatis. The arrow shows the slope to be measured later in QMAP (24° E, 17° N).

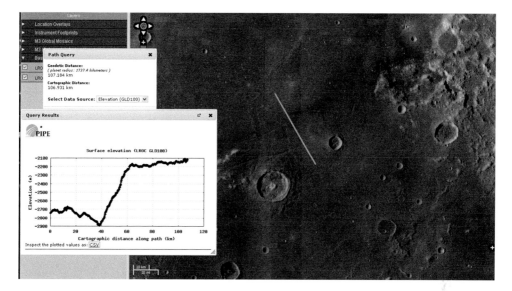

Figure 10.4 With QMAP, we draw a line (in green) to indicate the relief to be measured. The result is provided as a graph and series of values to be imported into a spreadsheet. The measurement shows a difference in altitude: Mare Serenitatis has subsided and formed grabens at the frontier of the mare, to the west of Promontorium Archerusia.

10.3 Identification

The classic way is using a printed lunar map or a printed lunar atlas. The most prominent is Antonín Rükl's *Atlas of the Moon* (Appendix 2). Until the breakthrough of webcam imaging, it was accurate enough to allow the identification of any lunar feature. As of 2016, however, we can see that, since the atlas was drawn by hand, it suffers a little from interpretation, and complex topographic features are somewhat hard to recognize with respect to modern, digital images. The renowned Virtual Moon Atlas software by Christian Legrand and Patrick Chevalley is very handy and accurate: it is far more suited to modern images. This is an invaluable tool. Its only weakness is that it does not take into account topography; hence features located very close to the poles, viewed edge-on, have a flat representation. For instance, the so-called Leibniz mons may be considered as mountains: they are rather crater rims, but this cannot be correctly represented without the assistance of a digital elevation model (DEM). Another, more advanced, software package is the Lunar Terminator Visualization Tool (LTVT), which is able to load a DEM from the Kaguya/Selene and LRO missions, allowing the correct identification of polar features.

11

Photogenic features of the Moon

11.1 Categories of lunar features

Table 11.1 lists the main features on the Moon, including official names in Latin and common names, in addition to some widespread notions. There is some confusion in naming because the origins of numerous features were (or still are) uncertain. For instance, Lacus and Palus refer to the appearance rather than a strict geological definition.

11.2 Maria

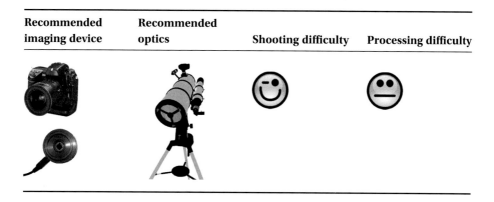

Recommended imaging device	Recommended optics	Shooting difficulty	Processing difficulty

Maria are supposed to contain only subtle topographic features (bowl craters, rimae, ridges, domes) and to exhibit limited variations in albedo, with the exception of possible pyroclastic deposits and ray systems. These places reveal surprises when the contrast is exaggerated (Figure 11.1): the highest levels may be substantially expanded beyond the mean soil brightness determined with the inflexion-point method. The increase in sharpness is less significant, and artifacts may quickly form in these regions of low levels, comparable to the camera noise unless numerous frames with relatively long exposures are taken. Maria are places where a photographer's creativity can be best expressed (Figures 11.2, 11.3, and 11.4).
    Recommended processing:

- threshold unsharp mask (Section 7.7.4);
- multiple-layer processing (Section 8.3) to expand brightness variations with no noise amplification while topographic features need some sharpness increase;

Table 11.1 List of lunar features and common terms used for the Moon

Common name	Latin name (if any)	Description
Concentric crater		More than fifty noticeable concentric craters exist; uncertain origin, possibly like crater chains but the impacts were aligned, or upwelling and then deflating lava, or internal rim mass wasting (e.g. Hesiodus A).
Crater		An excavation (rare small pits have a volcanic origin, most craters are caused by impacts), possibly encircled by a rim and, for younger and larger craters (>80 km), central peaks.
Crater chain	Catena	An alignment of small craters, possibly ejecta or successive impacts from a previously fragmented impactor (e.g. Catena Davy).
Craterlet		A small crater, less than 5 km in diameter.
Dome		A bulged volcanic feature, with or without vent. Often less than 4 km in diameter (e.g. west of Copernicus).
Ejecta		Matter (fused, pulverized rock, glass beads) blasted by impact, forming more or less extended bright rays.
Ghost crater		A buried crater covered by molten rock in a sea, only visible as a faint, softened rim when it is close to the terminator (e.g. Lamont).
Highland		An ancient, bright and raised terrain, heavily cratered, composed of anorthosite and coated by regolith.
Lake	Lacus	A small plain or small sea (e.g. Lacus Mortis).
Limb		The edge of the Moon (or any celestial body) as viewed from Earth.
Magcon		A buried MAGnetic CONcentration. There is only one on the near side: Reiner gamma.
Marsh or swamp	Palus	An intermediate terrain between a sea and a highland in terms of altitude and reflectivity. Palus Somni has an uneven terrain. Palus Epidemiarum is rather a small sea.
Mascon		A buried MASs CONcentration. A region in a mare where gravity is a little stronger, likely due to an underground volume of denser material. Invisible, but mapped because this affects the orbit of lunar probes.

Table 11.1 (cont.)

Common name	Latin name (if any)	Description
Mount, Mountains	Mons (single mount), Montes (mountain range)	A remnant of the rim of an ancient, huge crater (e.g. Montes Carpatus). The highest mount of the near side is Mons Huygens.[1] Sometimes flooded, with isolated peaks emerging from maria (e.g. Mons Pico).
Ocean	Oceanus	A very large sea, only one on the Moon: Oceanus Procellarum.
Plain	Planitia	Only one on the Moon, for historical reasons:[2] Planitia Descensus.
Promontory	Promontorium	A mountainous cape (e.g. Promontorium Archerusia).
Pyroclastics		Very dark, pyroclastic deposits of fragmented basalt with pyroxenes, olivine, titanium-rich beads, and glasses or quenched iron-bearing glasses (e.g. Taurus–Littrow).
Regolith		Extremely fine dust over all of the lunar surface, glueing because of static electricity (a small amount undergoes levitation during daytime because of the electrostatic charge due to the solar wind). Average thickness (variable) is 5 m on seas, twice as much on highlands. It results from meteorite impacts.
Rille	Rima (Rimae when it is a system)	Sinuous rilles are the remains of ancient lava flows (e.g. Hadley rille), or collapsed lava tubes (likely in the case of Rima Hyginus). Straight rilles are grabens, collapsed terrain between two faults (e.g. Rima Ariadaeus) resulting from a local elevation when magma rose underneath. Arcuate rilles are folded terrain due to the collapse of a sea, e.g. the center–East of Mare Humorum, also showing grabens at its edges.
Scarp	Rupes	Generally a fault, either linear (e.g. Rupes Recta) or curved (e.g. Rupes Altai).
Sea	Mare (plural Maria)	A large, dark, and shallow area, with small craters, formed of cooled, formerly upwelled magma coated by regolith, younger than highlands. Mainly the remnants of huge, ancient impacts (e.g. Mare Crisium).

[1] We do not take into account the unofficial Montes Leibnitz.

[2] The site of the first soft lunar landing by the Soviet Luna 9 probe in 1966.

Table 11.1 (cont.)

Common name	Latin name (if any)	Description
Secondary crater		A small crater, round or more or less elongated, not caused by direct impacts but resulting from fallout of soil debris blasted by an impact (e.g. North-east of Copernicus).
Gulf or bay	Sinus	Generally a tilted and partially buried, large ancient crater showing only an arc; the floor was flooded by a mare (e.g. Sinus Iridum).
Terminator		The boundary between the illuminated and the night part of the Moon (or any other moon, planet, or small body). Related to colongitude.
Valley	Vallis	A giant graben (U-shaped valley, when the crust subsides between two faults) or an ancient lava flow: sinuous rille or rima (e.g. Valles Alpis, which is both).
Wrinkle ridge	Dorsum (Dorsa when it is a system)	A ridge on a mare, due to surface folding when the mare subsided (e.g. Dorsa Smirnov in Mare Serenitatis).

- contrast and inflexion point (Section 7.6), unveiling rays (Section 8.4).

See also

- denoising (Section 7.7);
- tone mapping (Section 8.12).

The images should be taken very close to the full Moon to take advantage of the variation in albedo when the Sun strikes the surface at right angles. The purpose is not to detect subtleties of relief but to expand contrast.

11.3 Mountains

Recommended imaging device	Recommended optics	Shooting difficulty	Processing difficulty

Figure 11.1 A strong decrease of the lowest levels allows room within the dynamic scale to expand the intermediate and brightest levels in a mare. This results in a very contrasted image processed with the multiple-layer method, revealing ray systems and ejecta: one layer was moderately processed in sharpness with no consideration for the albedo, while the other layer was devoted to a strongly expanded luminance, then the layers were combined (28° W, 23° S).

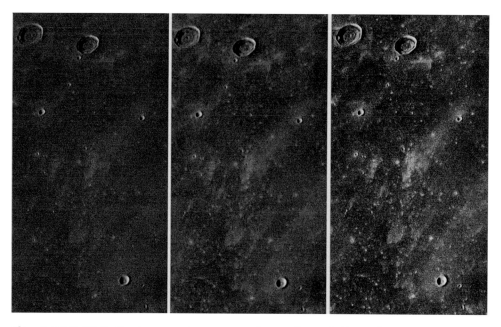

Figure 11.2 While the sharpness increase remains moderate (left), an abrupt expansion of levels just above the mean brightness of the surface reveals ray systems, ejecta, ancient lava flows, and some features emerging from the background (center). The images are enhanced with the inflexion-point method; tone mapping can be an alternative (right).

Figure 11.3 The single-layer processing results in more natural images. The interest of maria is underappreciated, but their wealth of detail is revealed by strongly contrasted images even if the increase in sharpness is more or less neglected (17° W, 36° N).

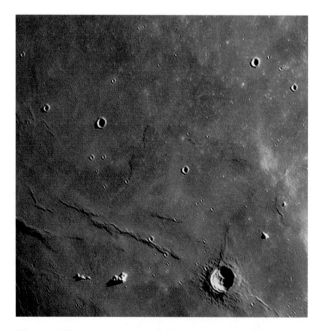

Figure 11.4 An unnammed, dark outpouring (lobate feature at the center) North of Lambert, revealed by the contrast adjustment rather than by its very discreet shadow. It was invisible even in the Virtual Moon Atlas (21° W, 25.7° N).

Figure 11.5 This image results from the use of several techniques: double-layer processing (one layer for crests and hills with wavelets; another one for the mare) with the inflexion-point method to emphasize the rays with little sharpness enhancement (application of a simpler, threshold unsharp mask is also a convenient technique, but this cannot reveal rays). The layers were fused and the image was eventually processed to balance the terminator gradient. Figure 9.6 shows the same region with the lighting reversed (0° E, 20° N).

These areas, most of which are located at the boundary between maria and highlands, show a very high dynamic range. Cameras with a large potential-well capacity are favored because they can simultaneously acquire the high levels of the crests and the low levels of the mare. The best moment for shooting a mountain chain is when the illumination is almost – but not exactly – vertical: the intrinsic contrast easily differentiates the mountain region and its surroundings (Figure 11.5). The maria prevent strong sharpness enhancement, while crests have a high signal/noise ratio; hence a threshold unsharp mask is often very useful.

Other dramatic views can be obtained when the crests cast elongated shadows. In such situations, the dynamic is limited and the well capacity has little impact. On the other hand, the levels are very low, demanding a low-noise camera with a very good analog-to-digital converter. A threshold in detectivity may cause posterization. Moreover, the level of the fixed-pattern noise is very similar to that of the signal, and little contrast enhancement is possible in these dark areas.

Figure 11.6 Near the terminator, the mountains are viewed edge-on. The processing consists of merely an increase in sharpness. The camera was rotated for framing, and then the image was cropped (6° E, 72° S).

Recommended processing:

- multiple-resolution unsharp mask (Section 7.5.2), threshold unsharp mask (Section 7.7.4), wavelets (Section 7.5.3);
- multiple-layer processing (Section 8.3) to expand brightness variations with no noise amplification on the maria, while topographic features need some sharpness increase;
- balancing the terminator gradient (Section 8.1) to enhance the visibility of the shadows cast on the gloomy area very close to the terminator.

Mountains and crater rims viewed edge-on at the limb (Figure 11.6) can be imaged at high magnification. Like the crews on the Apollo missions, we can observe that lunar mountains are rather rounded, resulting from fusion after huge impacts rather than plate tectonics (which is practically unknown on the Moon).

11.4 Cast shadows

Shadows cast by relatively young crater rims or crests of mountains chains provide emblematic views of lunar landscapes as they were depicted by artist astronomers such as Lucien Rudaux or James Nasmyth and James Carpenter in the days when lunar drawing offered better results than early film imaging. Photographing such shadows requires a low-noise camera and the stacking of numerous frames because the shadows are only slightly darker than the surroundings, but the scenes are worth the effort (Figures 11.7 and 11.8). Noise reduction is

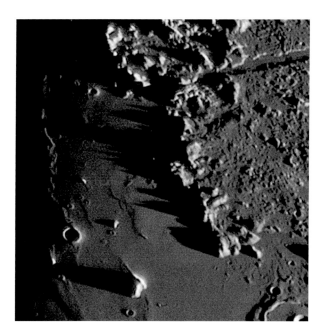

Figure 11.7 Shadows cast on a dark mare are poorly contrasted, while the image has a great dynamic; hence a few levels may be increased to enhance the contrast (1° W, 45° N).

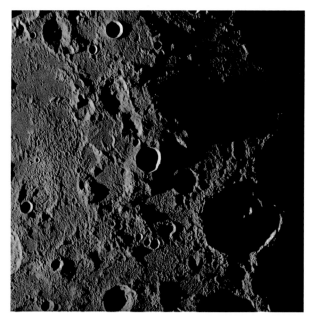

Figure 11.8 Shadows on highlands are easier to process than are those on maria (19° E, 10° S).

recommended if the shadows lie on a dark mare. The best locations are near the prime meridian, where shadows show their maximal elongation. In addition, the curvature of the Moon is detectable with the help of the brightness gradient on the crater floor, especially if the crater is more than 100 km wide and close to the lunar prime meridian.

Recommended processing:

- threshold unsharp mask (Section 7.7.4), wavelets (Section 7.5.3);
- balancing the terminator gradient (Section 8.1) to enhance the visibility of the cast shadows on the gloomy area very close to the terminator.

See also

- denoising (Section 7.7);
- time-lapse (Section 11.6).

11.5 Highlands

Recommended imaging device	Recommended optics	Shooting difficulty	Processing difficulty
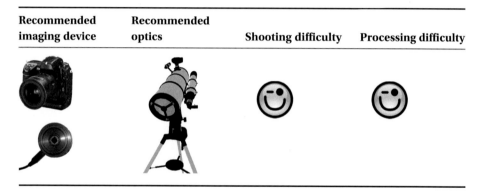			

Highlands have roughly twice the albedo of maria (0.12–0.18 vs. 0.07–0.1), and the exposure needed is shorter. Since they do not show the same extreme dynamic as maria, the average level is much higher and the signal/noise ratio is very favorable (Figure 11.9). As a consequence, highlands tolerate stronger sharpness processing (Figure 11.10).

Recommended processing:

- wavelets (Section 7.5.3) or threshold unsharp mask (Section 7.7.4).

See also

- detecting rebounds (Section 8.7);
- tone mapping (Section 8.12) and multiple-layer processing (Section 8.3).

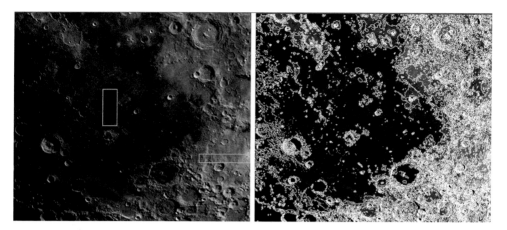

Figure 11.9 Left: an image is framed on Mare Nubium close to a highland terrain. Right: the isophote (IRIS, Fitswork . . .) or topography (PaintShopPro™ . . .) representation shows a wider range of levels in highlands, meaning a better signal/noise ratio and more freedom for tone mapping and sharpness enhancement. The mare (yellow frame) contains 4000 distinct levels only, all being close to the noise level, while the highland (blue frame) contains 45 000 distinct levels (12° W, 22° S).

Figure 11.10 Highlands are often the easiest images to process; they can take a strong sharpness enhancement (wavelets, unsharp mask . . .) and need little contrast correction (3° W, 29° S).

267

11.6 Sunrise and sunset in craters

Recommended imaging device	Recommended optics	Shooting difficulty	Processing difficulty

The terminator shifts by 15 km per hour (9 miles per hour) at the equator. The scene evolves in hours (Figures 11.11 and 11.12), and image sequences or time-lapse sequences are interesting at high magnification with automated reframing using PIPP or Dave's Video Stabilizer (Appendix 2) to suppress image jerking. The evolution over a period of some days is also interesting. The best

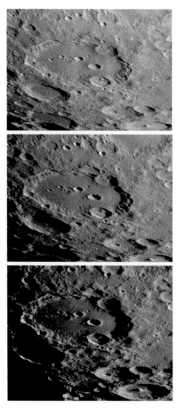

Figure 11.11 Effect of the terminator shifting near Clavius. Note the brightness gradient on the crater floor (in the bottom image), due to the sphericity of the Moon (15° W, 59° S).

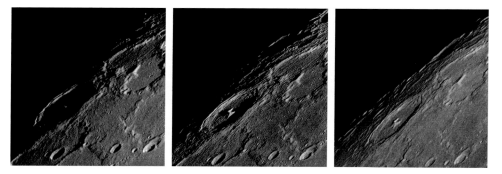

Figure 11.12 The complex and arcuate central peak of Pythagoras gradually emerges from darkness (62° W, 64° N).

subjects are relatively large craters with a central peak: the shadow of the peak is cast on the internal rim, rises, and then masks it, while the base of the peak plunges into the absolute darkness of the shadow cast by the opposite rim.

Recommended processing:

- wavelets (Section 7.5.3), threshold unsharp mask (Section 7.7.4), multiple-resolution unsharp mask (Section 7.5.2);
- balancing the terminator gradient (Section 8.1), emphasizing wrinkle ridges (Section 8.6) to enhance the visibility of the casted shadows on the gloomy area very close to the terminator.

See also

- local contrast (Section 8.2).

11.7 Phase effects

Recommended imaging device	Recommended optics	Shooting difficulty	Processing difficulty
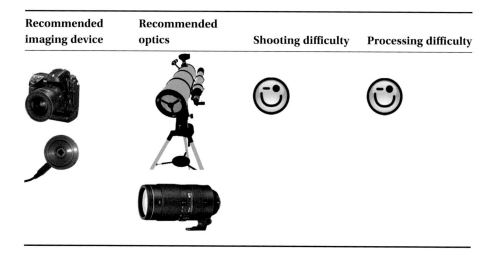			

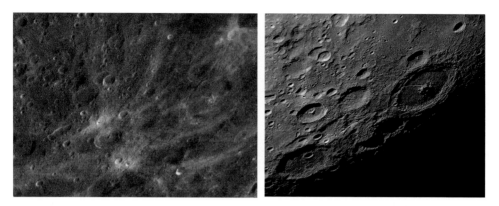

Figure 11.13 Petavius (at right) may be barely identifiable when it is far from the terminator. Apart from the topography, the variation of lighting conveys additional information about the reflectivity – and age – of the different craters. The small craters Stevinus A, Furnerius A, and Furnerius C turn into lighthouses at full Moon, eclipsing all of the larger neighboring craters (56° E, 28° S).

The differential variation in albedo of various areas and features is easily noticeable when images are taken over an interval of several days as the phase varies (Figures 11.13 and 11.14). This requires a medium or long focal length and a relatively wide field, hence a DSLR or a megapixel planetary camera is suited to the subject, with a powerful telelens, or a telescope at prime focus or with a limited magnification. The ideal field of view is about 500 km wide, representing a correct balance to see medium-resolution topographic details and large-scale features like ray systems, especially around Copernicus or Tycho.

Recommended processing:

- multiple-resolution unsharp mask (Section 7.5.2), wavelets (Section 7.5.3);
- multiple-layer processing (Section 8.3), contrast and inflexion point (Section 7.6), unveiling rays (Section 8.4).

See also

- tone mapping (Section 8.12).

Another consequence of the lunar phase is the inversion of the illumination angle after two weeks. This provides indications about the real slope of asymmetrical reliefs.

11.8 Grabens, linear rilles, rifts

Recommended imaging device	Recommended optics	Shooting difficulty	Processing difficulty

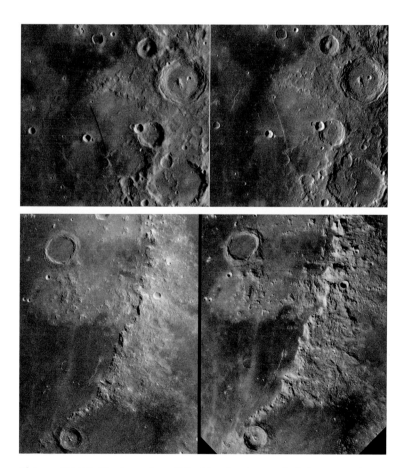

Figure 11.14 The inversion of the illumination angle brings complementary views of the topography (left: 6° W, 21° S; right: 5° W, 21° N).

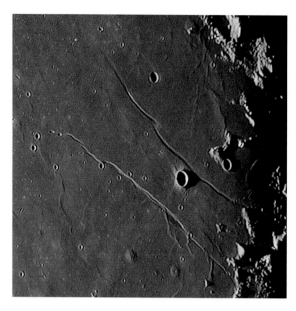

Figure 11.15 The Cauchy region (38° E, 9° N) hosts a rift (South), whose slope is strongly lit because it faces the Sun, and a graben (North) with a bright slope and a dark, opposite slope. It requires a low angle of illumination while shooting. Since the region is very dark and the lighting was tangential, the image was brightened with a Shadow/Highlight function (37° W, 10° N).

If the sunlight makes an acute angle with the surface and is conveniently orientated relative to topographic features, the Sun strongly lights up the facing, rectilinear scarps, while opposite scarps are in shadow. Rilles in fractured terrains, ancient lava channels, and grabens show both lit and shadowed areas, because they are U-shaped (Figures 11.15 and 11.16). This requires at least a 125-mm (5-in) telescope because the rilles are often 300 m deep and only 1.5–2 km wide, although they can extend to some hundreds of kilometers in length. Rifts are rarer but prominent, for instance Rupes Altai (Figure 11.17) and Rupes Recta.

Recommended processing:

- threshold unsharp mask because most rilles are located on weak-signal/- noise-ratio, dark maria (Section 7.7.4), multiple-resolution unsharp mask (Section 7.5.2), wavelets (Section 7.5.3).

See also

- balancing the terminator gradient for sinuous rilles at low colongitude (Section 8.1);

Figure 11.16 The rilles in the fractured floor of Gassendi show a surprising variety of shapes. They can easily be shot under relatively high-angle lighting. Moreover, the crater floor is rather bright and rough despite the ancient, invading flood from the nearby mare. The region also shows numerous and concentric grabens and wrinkle ridges at a large scale, centered on Mare Humorum (39° W, 17° S).

- denoising (Section 7.7), because rilles are superimposed on maria, and these equally dark features lead to a poor signal/noise ratio.

11.9 Valleys

Since the rare great valleys are located on highland-type terrains, image processing applicable to highlands is applicable here (Figures 11.18 and 11.19). The signal/noise ratio is very favorable, and they do not need a high magnification: a simple beginner's refractor is sufficient, as long as the lighting is favorable. Numerous grooves due to huge ejecta which have abraded the region around Albategnius are other splendid subjects; they may be processed in the same way.

Recommended processing:

- multiple-resolution unsharp mask (Section 7.5.2), wavelets (Section 7.5.3).

11.10 Secondary craters, chain craters

Recommended imaging device	Recommended optics	Shooting difficulty	Processing difficulty

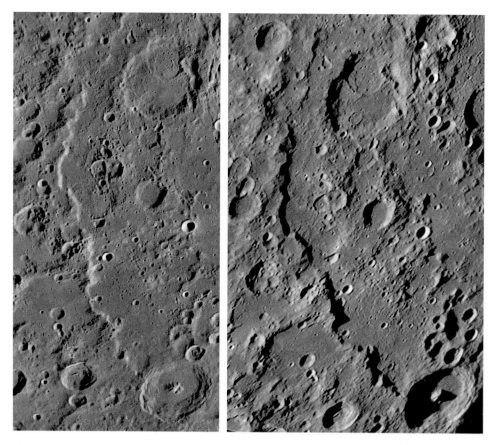

Figure 11.17 Rupes Altai (23° E, 24° S) is an easy target for all telescopes, though the area is not very well contrasted. The feature should be processed like a giant crater rim, essentially in sharpness enhancement while keeping a contrast margin to avoid clipping of high levels (26° E, 23° S).

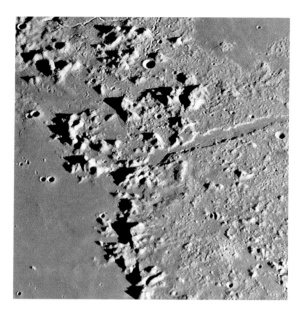

Figure 11.18 Vallis Alpes is a giant graben split by a wandering, ancient lava channel. Both show a well-lit slope and a darkened one. Shooting the channel requires a well-collimated 180-mm (7-in) reflector and an acute angle of illumination, with classic sharpening processing (3° E, 48° N).

Figure 11.19 The ruined Janssen walled plain crater has three delicate rilles (Rimae Janssen), which reveal themselves under grazing-light conditions. Vallis Rheita, the vertically orientated, long valley, is visible only under comparable conditions, but it is an easy target for small instruments (44° E, 41° S).

These small features (Figures 11.20 and 11.21) require a stable atmosphere and a well-collimated, large-aperture telescope. Low-angle illumination is recommended, less than 20° or so relative to the colongitude. The contrast is not a concern, while the emphasis should be placed on the sharpness.

Figure 11.20 Secondary craters east of Copernicus. The craters are located on a maria-like area, and a threshold unsharp mask preserves the image from increasing noise. Bottom right: the Stadius ghost crater barely emerges from the surface (15° W, 13° N).

Figure 11.21 The Catena Abulfeda crater chain is more easily processed because it stands on highlands, with a better signal/noise ratio. The trap to be avoided is confusing the topographic details with artifacts from too strong a sharpness processing (16° E, 16° S).

Recommended processing:

- threshold unsharp mask, if the secondary craters lie on a mare (Section 7.7.4), multiple-resolution unsharp mask (Section 7.5.2), or wavelets (Section 7.5.3).

See also

- denoising, if the secondary craters lie on a mare (Section 7.7).

11.11 Pyroclastics

These are the very darkest features of the Moon. They show a weak signal/noise ratio, no relief, and few details, but they become fascinating when low levels are stretched while high levels are compressed. Vertical lighting strongly differentiates the albedo of pyroclastics with respect to the surroundings (Figures 11.22, 11.23, and 11.24). The sharpness is not very important, because small-scale details convey no further significant information.

Recommended processing:

- threshold unsharp mask (Section 7.7.4), multiple-resolution unsharp mask (Section 7.5.2), or wavelets (Section 7.5.3);
- inflexion point (Section 7.6);
- multiple-layer processing (Section 8.3) to manage both contrast and sharpness;

Figure 11.22 Taurus–Littrow lies at the boundary between a mare and a highland. The lowest levels were set to minimum and then a small interval of low levels was expanded, resulting in a strong contrast in the pyroclastics (almost black) and more conspicuous differences in the soil of the two neighboring maria. The brightest levels were both increased and compressed, causing an inevitably lesser contrast in crater rims, scarps, and rilles: this is the price of the legibility of the feature (28° E, 18° N).

Figure 11.23 Sinus Aestuum is darker than Mare Vaporum. The contrast was adjusted with an S-shaped curve, expanding low levels and compressing high levels (3° W, 14° N).

Figure 11.24 Most pyroclastic deposits in Alphonsus (three large ones and at least two smaller ones) stand on sinous rilles. The contrast was enhanced with the inflexion-point method and the sharpness was improved with a threshold unsharp mask (13° W, 13° S).

- denoising, because pyroclastic deposits have mainly flooded onto maria, and both features are very dark (Section 7.7).

See also

- tone mapping (Section 8.12).

11.12 Magcon

There is only one magcon located on the near side (Figure 11.25). It reveals no relief at all, but, owing to the supposed action of its measured, small magneto-sphere, it protects the soil from the darkening effect of the solar wind. This results in a strong local albedo, easily differentiated from the surrounding maria, and hence the best period to shoot it is when its location relation to the colongitude is maximal.

Recommended processing:

- threshold unsharp mask (Section 7.7.4);
- inflexion point (Section 7.6);
- balancing the terminator gradient (Section 8.1);
- multiple-layer processing (Section 8.3) to manage both contrast and sharpness, or unveiling ray systems (Section 8.4), because the technique is useful here.

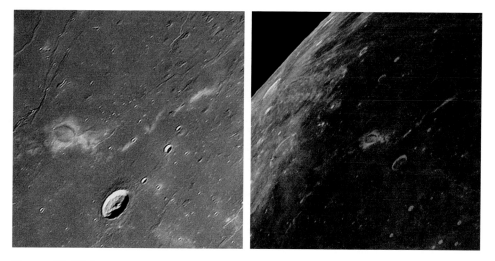

Figure 11.25 Reiner gamma, near the Marius crater, is the only magcon located on the near side. The best contrast is obtained when it is far from the terminator, but, because its albedo is strong while it is located in a dark mare, a difference of 15–20° relative to the colongitude is sufficient to reveal it. The inflexion-point method helps to differentiate the magcon and its surroundings (59 °W, 7° N).

See also

• tone mapping (Section 8.12).

11.13 Domes, ghost craters, and ridges

Recommended imaging device	Recommended optics	Shooting difficulty	Processing difficulty
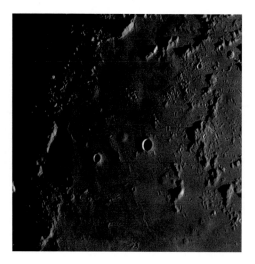			

These are among the most elusive features of the Moon. Not only they are as dark as the maria on which they stand, but also they are small – domes are about 5 km in width and domes (Figures 11.26 and 11.27), ghost craters, and ridges (Figure 11.28) are some hundreds of meters in height. They demand a very low angle of illumination and a substantial stacking, e.g. 500–1000 frames, and a low-noise camera. Even if 8-bit captures suffice, it is essential to process such images with 16-bit (or more) image editors because the lowest

Figure 11.26 Domes always show a low dynamic range and the levels are close to the noise. Left: South-west of Copernicus. A grazing lighting angle is required in order to reveal them (29° W, 10° N).

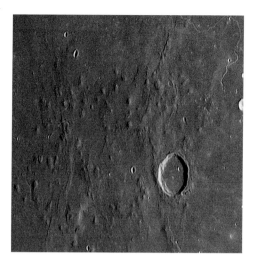

Figure 11.27 West of Marius. A strong brightness-gradient adjustment was performed. The images were taken with a noisy sensor, requiring an increased number of frames to improve the signal/ noise ratio; nonetheless, some bands, posterization, and roughness remain detectable. Low levels were compressed so that the background noise level became similar to the lowest significant signal, then the image was denoised (52° W, 13° N).

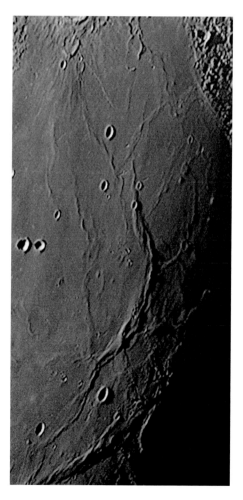

Figure 11.28 Being visible only when they are close to the terminator, images of wrinkle ridges are prone to the same flaws as images of domes and ghost craters. Fortunately, a moderate magnification is sufficient, allowing a better contrast. The main processing here was devoted to mitigating the gradient (53° E, 3° S).

levels are poorly differentiated; use of 8-bit editors could result in unfixable, visible posterization.

Recommended processing:

- threshold unsharp mask (Section 7.7.4);
- balancing the terminator gradient (Section 8.1).

See also

- tone mapping (Section 8.12);
- denoising (Section 7.7).

An example of a ghost crater is shown in Figure 8.10.

11.14 Libration

Recommended imaging device	Recommended optics	Shooting difficulty	Processing difficulty

Libration (Figure 11.29; see also Figure 1.14) is a relatively wide-field subject, and it is visible at almost all lighting angles; it primarily demands patience. Lunar ephemerides help one to select the right dates for shooting. No particular processing is required; the only minor difficulty is the height of the Moon, because thin crescents are never far from the Sun, and hence never high in the sky before sunrise or after sunset: variations in accuracy are to be expected.

Recommended processing:

- multiple-resolution unsharp mask (Section 7.5.2) or wavelets (Section 7.5.3);
- inflexion point (Section 7.6) to emphasize a possible mare close to the limb.

See also

- making 3D images (Chapter 9).

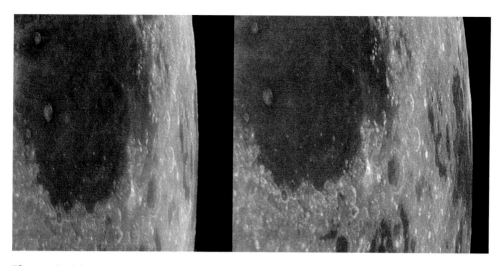

Figure 11.29 Lunar libration allows us to see 59% of the Moon. This is illustrated in Section 1.2.5 and in this image, showing Mare Marginis at the limb. A simple contrast expansion easily differentiates the highlands and the mare (66° E, 12° N).

11.15 Monitoring transient lunar phenomena (TLP)

Recommended imaging device	Recommended optics	Shooting difficulty	Processing difficulty
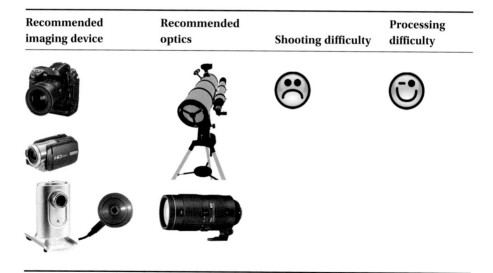			

These are poorly understood, short-lived (seconds to hours) changes in brightness, mostly reported as spots or fuzzy, small areas, possibly with colors. Some happen because of meteoroids, related to meteor showers (e.g. the Leonids).

Others could arise from electrostatic phenomena, such as regolith material undergoing electrostatic levitation because of daily – from a lunar point of view – variations in the solar wind; escape of gases, perhaps when the Moon is at its smallest distance from the Earth, which is when its gravitational interaction is strongest, causing outgassing from cracks; or other perturbations that could happen fairly frequently, such as small moonquakes. The Apollo 11 and Apollo 15 crews, along with professional astronomers, noticed unexplained, ephemeral changes, and a spectrum[3] has even been taken, showing evidence of molecular emission lines.[4] Radon emissions have been reported by lunar missions.

The most interesting sites are Aristarchus, Plato, Schröter's Valley, Mare Crisium, Proclus, Alphonsus, Herodotus, Gassendi, Agrippa, Tycho, Kepler, Grimaldi, Copernicus, and Ross D (ordered approximately in descending number of reported TLP according to the publications). Aristarchus is undoubtly the most promising place to monitor.

A photographic survey program can be undertaken with the inspiration of the joint program by NASA's Meteoroid Environment Office and Marshall Space Flight Center's Space Environments Team. In 2005–2006, eleven meteoroids were recorded with two sets of CCD video cameras and 14-in SCT telescopes (Figure 11.30). In 2014–2015, twenty-five impacts were recorded in twelve months with the same equipment. Some impacts reached the eighth and ninth magnitudes, due to modest, 2–3-in, some-hundred-gram impactors. Such a survey cannot be conducted when the Moon is gibbous or full because the apparent magnitude of the full Moon is –12.7, to be compared with the eighth–ninth magnitudes of the aforementioned events.

Another very affordable possibility is a wide-angle survey of the entire lunar hemisphere with a small sensor camera and a photolens to detect meteoroids; this also requires a motorized mount set for lunar tracking. Image-acquisition software for time-lapse photography (e.g. FireCapture, Gstar Capture 4) and a video/deep-sky/planetary CCD camera are perfect for this purpose. Cameras are sensitive to electromagnetic glitches and cosmic rays, and, for the latter, the probability increases with altitude. Cosmic rays are easily recognizable because they form perfectly defined, short, white lines or dots on raw frames. The best remedy is to duplicate the imaging system, as professionals do, with two sets of imaging equipment on a plate attached to the mount by the dovetail, to ensure that the event is not due to a glitch in one imaging device. The strategy of doubling the equipment to

---

[3] An amateur spectral survey could be undertaken with a sturdy mount, reliable lunar tracking, and a spectrograph such as Shelyak Instruments's Alpy-600, but this is verging on professional equipment.

[4] http://adsabs.harvard.edu//full/seri/PASP./0071//0000046.000.html

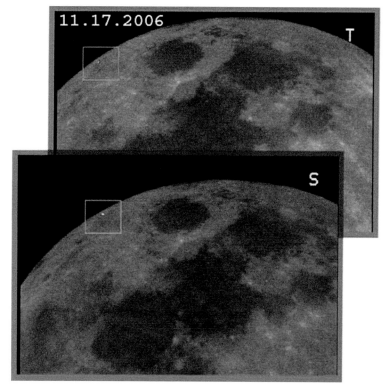

Figure 11.30 The same impact recorded with two 14-in SCT and video cameras on November 17, 2006 at the Automated Lunar and Meteor Observatory. Image by NASA.

eliminate measurement errors is already being applied by amateurs (to measure transits of exoplanets); two members of an astronomy club could undertake such a survey, even if their equipment is not exactly the same. Moreover, a satellite could cross the Moon and twinkle if it rotates and reflects the Sun. To eliminate such an intruder, we can either consult www.heavens-above.com (a must-see website for satellite tracking and recognition) or monitor the Moon from different locations in latitude and/or longitude. Of course, aircraft may cross the Moon too.

The frames must be studied with absolutely no processing in order to avoid introducing artifacts; this is why CCD sensors are best suited, because the image is very clean (e.g. Sony's ICX or Kodak's KAI/KAF sensor family). As long as the location, the precise time, and the filter used, if any, are mentioned, any photography or visual observation could be useful. The best advice is to alert ALPO (Appendix 2), if the report is serious. Of course, such a report is an indication rather than constituting proof, and countless observations of possible TLP have been reported with caution by experienced professional and amateur astronomers because of the possibility of misinterpretation of simple variations in seeing due to turbulence.

See also

http://user.astro.columbia.edu/~arlin/TLP/ by Arlin Crotts (University of Columbia)
http://nasasearch.nasa.gov/search?utf8=%E2%9C%93&affiliate=nasa&query=
lunar+transient+phenomena (TLP reports at NASA)

12

Naming, archiving, printing, and sharing lunar images

12.1 Selenographic, cartesian, and xi–eta coordinates

Selenographic coordinates are useful to locate features and to name image files. These are expressed as latitude and longitude, like on Earth. A standard from the International Astronomical Union (IAU) defines East as the direction of rotation of a planet[1] or a moon. This is not obvious for the Moon because it is in synchronous rotation with the Earth, a common behavior for a moon.[2] The only way to detect the rotation of the Moon, viewed from the Earth, is to observe the phase. If the lighting comes from the Mare Crisium side, this shows the lunar East. This is a valuable marker whichever hemisphere of the Earth we observe the Moon from. Viewed from the northern hemisphere, the lunar East is on our right hand; from the southern hemisphere it is on our left hand. The lunar North is at the top when viewed from the northern hemisphere, and at the bottom when viewed from the southern hemisphere. Like on Earth, latitudes and longitudes are measured in degrees, from 0 to 360 (Figure 12.1). The East is at +90° in longitude, or 90° E. The North is at +90° in latitude, or 90° N. West and South have negative coordinates. The prime meridian crosses the equator in the Sinus Medii, where there is absolutely no remarkable feature but the little, 3-km (2-mile) Oppolzer A crater (accessible to all telescopes) at 0.345° W and 0.485° S. Since the Moon is almost spherical, the coordinate system is based on trigonometry.

A simpler coordinate system is the cartesian diagram. It consists of splitting the Moon into squares or rectangles. The drawback is that this system is not very widely used, with the noticeable exceptions of Antonín Rükl's highly acclaimed *Atlas of the Moon* and the Lunar and Planetary Laboratory's System of Lunar Craters.[3] Rükl's system maps the Moon in seventy-six squares;[4] numerous atlases and books intended for amateurs still use this system – in addition to

---

[1] Venus and Uranus rotate westward, or, in other words, they rotate eastward but upside down.

[2] Several moons are both in synchronous rotation and in synchronous revolution with their primary, so that not only does the same side always face the planet, but in addition the moons are always located at the same point in the sky, viewed from the planet. The most spectacular case is Pluto's moon Charon.

[3] https://the-moon.wikispaces.com/System±of±Lunar±Craters

[4] https://the-moon.wikispaces.com/Rukl±Index±Map

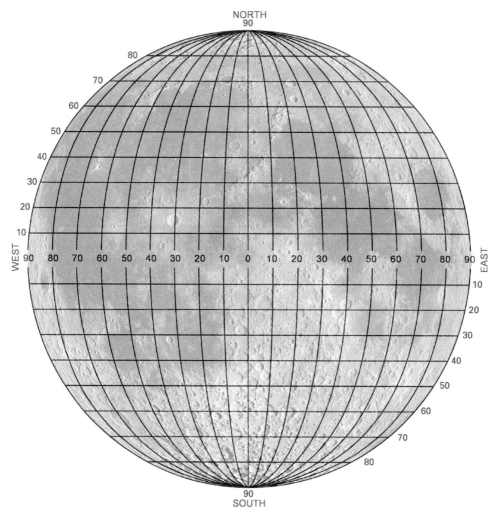

Figure 12.1 The IAU coordinate system.

selenographic coordinates – because it is commensurate with the field of view of small-aperture telescopes, and it does not require any particular ability in mental arithmetic to instantly calculate trigonometric conversions (Figure 12.2).

A convenient way to convert selenographic to cartesian coordinates is the xi–eta coordinate system (Figure 12.3). Xi is the horizontal coordinate starting from –1 (lunar West) to +1 (East). Eta is the vertical coordinate, from +1 (North) to –1 (South). The scale is linear. Since the 0,0 xi–eta coordinates correspond to the 0,0 selenographic coordinates, the purpose is to convert the coordinates from one system to the other as follows:

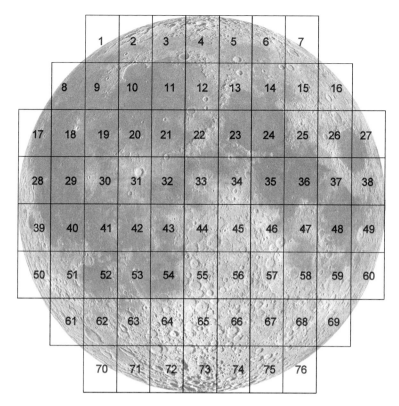

Figure 12.2 Antonín Rükl's coordinate system.

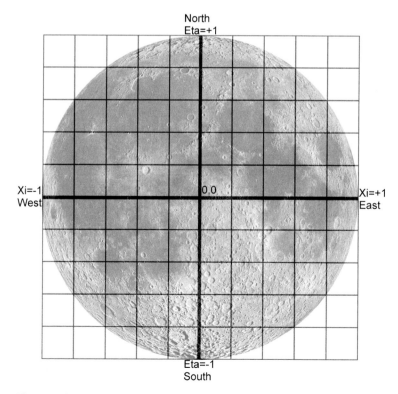

Figure 12.3 The xi–eta coordinate system.

Latitude = arcsin(Eta);
Longitude = arcsin(Xi/cos(Latitude))

The reverse conversion is

Xi = cos(Latitude) × sin(Longitude)
Eta = sin(Latitude)

The calculation can be performed by a calculator or by spreadsheet software. Thus, the simple cartesian coordinate system can be expoited for targeting and for naming image files, then the coordinates may be converted at any time into selenographic coordinates.

12.2 Image-file naming

The first idea is to name each image according to the main features it shows, for instance large craters or a mare:

Gassendi.tif

The author takes some 150–200 decent images of the Moon per year (depending on the weather!), so manually naming the image quickly proved to be a tedious task. Another solution is naming images according to lunar coordinates, or selenographic coordinates:

40w17s.tif

The accurate coordinates are given by software. Rounded coordinates are sufficient, with no decimals. One of the most appropriate items of software is the Virtual Moon Atlas by Christian Legrand and Patrick Chevalley (Appendix 2). The information tab gives the exact coordinates once we have clicked on the feature in the map. Another way is to position the mouse on the map with respect to the center of the image. The status bar at the bottom indicates the coordinates.
If we use Rükl's coordinates, the image-file naming convention may become

52.tif

If the technical details are constant (same telescope, same camera, same magnification . . .), a simple text file can mention them for all images. If they vary, the image names can include an abbreviated description of the equipment and shooting conditions, using for instance underscores (_) to separate data:

40w17s_12in-meade-sct_dmk31_zeiss-barlow-2x_exp-15 ms_gain-150_seeing-2.tif

Another solution is to adopt a code for technical information, with a short reference to a typical setup in the image name, and a common text file listing the codes and the corresponding information:

40w17s_setup01_exp-15 ms_gain-150_seeing-2.tif

The text file common to all images:

setup01 12" Meade SCT, DMK31 camera, Zeiss Barlow 2x
setup02 5" Celestron SCT, Skynyx 2.0 camera, Skywatcher ED Barlow 2x

(and so on).

If we hope our images will be referenced by Web search engines, the best thing to do is to name the image with human-readable information, that is, mention the main target of the image and the abbreviated setup.

Feature-orientated name:

Gassendi_8in-newton_skynyx2,0.tif

Selenographic-coordinates-orientated name:

40w17s_8in-newton_skynyx2,0.tif

Or Rükl's-coordinates-orientated name:

Rukl-52_8in-newton_skynyx2,0.tif

The final image may include a full description in a sidebar. In addition, if the image is included in an HTML page, the page may contain a full-length, human-readable description of the setup with the names of the main features of the image and one or more coordinate systems. The most important keys to search for a lunar image, for amateur astronomers, are the feature, the instrument, and the camera. The feature is the main subject. The telescope and camera can be used to compare other amateurs' images with our own work done with comparable equipment.

Of course, we can image the same target several times with the same equipment. Therefore, the date should be mentioned, not only to ensure that each image file will have a unique name, but also to keep information about the libration and the height of the Moon. These data may be retrieved afterwards with the help of software packages such as Virtual Moon Atlas, Stellarium, and others. Given the fact that the coordinates or the feature are the main criteria for a search, the date can be the secondary criterion because we are expecting to improve our images with experience. The time stamp for astronomical images should be compliant with respect to Universal Time (UT) because it is independent from the location. We can take advantage of the fact that the operating system/file manager automatically sorts the image files by name and adopt a simplification of the ISO 8601 standard, so a good habit is to write a date as YYYYMMDD-HHMM. The files will be automatically sorted by feature/coordinates, then by date and time. With the exception of some particular subjects such as occultations, eclipses, conjunctions, or time-lapse sequences, the hour and the minutes may be neglected in most cases, unless the Moon was imaged when it was very low above the horizon.

Finally, among numerous possibilities, the author has adopted a convention for Web-published images with a human-readable name, such as

moon-35w15s-gassendi_2013-02-21_newt254_asi120 mm.jpg

Original movies (SER, AVI) are named with the same rules. The magnification (Barlow lens or eyepiece projection) and the filter are omitted from the file name, but the processed image contains all of the technical information in a sidebar. AVI movies have no containers like EXIF data or a FITS header, but SER movies do. Acquisition software such as FireCapture may create an associated TXT file containing stated acquisition data.

Other information such as the camera exposure and gain, seeing conditions, and filters can be embedded in the image, depending on the file type. EXIF data and a FITS header are convenient ways to store additional information. But we have to be careful because some software may lose, affect, or rewrite these items of information. Another precaution is to create a simple text file with the same name and a different extension (.TXT) to store additional information. If we create webpages, this text file may be automatically merged with the HTML file with the help of parsing software (it reads the content of the .TXT file then translates it into HTML and generates a <title> tag). Preserving shooting conditions is especially important in petrographic imaging: the filters and their bandpass have to be clearly mentioned, ideally with a description of the processing.

Note: a good precaution is to use no more than sixty-four characters and authorized symbols in order to be compliant with the Joliet/ISO 9660 standard for naming files to be written onto a CD-ROM. The standard is applied by numerous operating systems and CD-burning software packages, but we have to take care of the default parameters prior to burning the CD.

12.3 Archive management

12.3.1 CD, DVD, Blu-ray

CDr/rw and DVDr/rw (Table 12.1) have proved not to be lasting archives for long-term storage. Even choosing major-brand discs is not always a guarantee

Table 12.1 Capacity of common, 12-cm optical discs. This is to be compared with the average size of lunar movies or images. An AVI may be compressed by more than 50%, but a SER movie accepts little compression.

CD	DVD	Blu-ray
700 megabytes	4.7 gigabytes (single layer)	27 gigabytes (single layer)
	8.5 gigabytes (dual layer)	50 gigabytes (dual layer)
		100 gigabytes (triple layer, BD-XL)
		128 gigabytes (quad layer, BD-XL)

because of unmanageable fluctuations in manufacturing processes from subcontractors. In particular, the opaque layer (golden or silver) has to be preserved from exposure to daylight. When stored in good conditions, with no scratches and away from light, we can count on a four- or five-year durability for CDs/DVDs.

Professionals are used to managing "refreshment" of archives: the data are copied onto another, new disc each year. This is possible when we have a limited number of discs. Of course, archiving once per year is risky. Imaging the Moon is a continuous activity, and using rewritable discs seems more secure. But we have to keep in mind the following facts.

- Simplified disc-burning software does not replace old data by new data, but, instead, accumulates new data in addition to old (henceforth invisible) data.
- Erasing the CD/DVD is more reliable, but this process is tedious when we frequently archive images. Furthermore, there is a large but still limited number of rewriting operations (several hundred at least).
- We have to be aware of the default settings of the software: sometimes they prevent one from carrying out multiple rewriting operations.
- A CD/DVD lacks the room to store numerous movies. Each movie can occupy up to one or two gigabytes, which is more than an entire CD. A solution is to save only stacked images rather than cumbersome movie files, assuming that the registration/stacking process was as good as possible. As the stacked image is the basis on which to try using different types of processing, it is safe to archive it. In previous years, great differences in results between software packages prevented you from choosing your favorite software giving "definitive" stacked images. As of 2016, stacking software is mature enough to help us to make a choice based on the balance between the results and the ease of use, and we may consider that the movie can be discarded after the stacked image has been archived.

Since CDs/DVDs/Blu-rays are cheap, a good precaution is to make duplicates to be entrusted to someone in another location. It is reckless to store the original discs in the same room as the computer.

12.3.2 USB keys, USB disks

Removable, solid-state media have, as of 2016, barely the same capacity as hard disks. Since the goal is archiving rather than copying every day's work, we can neglect the transfer speed because saving can be initiated at night. In the case of a possible electric disaster while data are being transferred with no supervision, a rotation between two media is an excellent precaution. CDs/DVDs/Blu-rays are sensitive to scratches, some types of fungus, daylight, mechanical stress, and time. USB disks and USB keys are sensitive to electric, magnetic, and electromagnetic influences. USB keys and USB SSDs (solid-state disks) are insensitive to shocks. The capacity greatly exceeds the biggest quadruple-layer

BD-XL (128 megabytes). The main limit is the transfer speed (as of 2016, USB3 reaches at least 70–145 megabytes per second and S-ATA II more than 550 megabytes per second). Owing to the lack of years of experience, the durability remains unknown for SSDs, but even cheap USB keys last much more than five years – longer than optical discs – with a much greater ease of use than optical discs. However, limited capacity and greater cost limit the number of movies which can be archived. The author acquires movies on a laptop, stacks images, then transfers the stacked images with a USB key to a desktop for processing, and ensures that the stacking is correct,[5] prior to erasing the movies from the laptop to free room on the disk.

12.3.3 Cloud storage

Cloud storage via the Internet is another good solution for archiving. There are some drawbacks to this attractive solution. In order to offer a simple and pleasant interface, some applications require the installation of a Java engine or an executable appliance. Both are resident and consume a small amount of one's computing and memory resources; they bring about a loss of performance for image acquisition at high speed and constitute a potential security flaw. We may feel reluctant to scatter images somewhere on the Web, even through a VPN (virtual private network), even if the provider guarantees privacy (servers are often shared resources on multiple-customer platforms). Cloud storage was primarily intended as a means to access files via different portable devices at the price of limited file size and variable bandwidth, especially WiFi or Bluetooth, but, even at the time of writing, many people have a limited bandwidth, leading to a poor transfer rate that is incompatible with the size of a movie. The capacity depends on the provider and subscription; it is at least (in 2016) 1–2 GB: enough to store stacked and processed images, but not enough for movies. On the other hand, cloud storage is perfect to share stacked images, for instance in order to compare different types of processing of the same image by the scattered members of an astronomy club. Unfortunately, we cannot list cloud storage providers at this time because the market is fluctuating, but things are constantly evolving in a favorable way.

12.4 Offset printing and pre-processing

Publishing images in an amateur astronomy magazine is a thrilling experience . . . until we look at the result. This is due not to the editorial team but to offset

---

[5] Sometimes, in poor seeing conditions, the alignment points or the quality parameter prove to have been badly chosen. A corrective action is performed, then, after the stacked images have been transfered to the desktop and then verified, the movies are erased from the laptop. The rule is to always keep two copies of the stacked images (or the possibility to re-generate them) on two separate disks.

printing, which alters images. Fluctuations in materials and processes lead to variations even for individual copies of the same issue of the magazine. Moreover, astronomical images demand particular attention. Ordinary pictures are often printed with an intensified density, that is, with the low levels lowered to enhance the colors. Lunar (and deep-sky) images contain a large proportion of *dark* areas that are not *black* areas, especially maria or domes near the terminator. Furthermore, it is impossible to print deep black in a magazine, quite the opposite of a photography book. This is why images to be published in a magazine have to be processed in a special way.

- The level scale must be adapted: it is better to preserve a large number of levels for the darkest and brightest levels, while medium tones (highlands and maria illuminated almost at a right angle) have to be compressed around intermediate values.
- A chart can be sent accompanying the image to help the magazine's staff to calibrate the levels. The author successfully shared a chart with a magazine (*Astronomie Magazine*) in order to enhance the rendering of lunar images.
- The very smallest details are often drowned by the offset-printing pattern (a kind of network of dots simulating plain surfaces at a distance). If the image is devoted to a given feature, it is safe to crop the image in order to maintain the details at a certain scale above the possible pattern. If the image is to present a large area, tiny details will disappear.
- The best advice is to lighten the surface with great margins for the darkest (e.g. domes and wrinkle ridges) and lightest (e.g. terraces in the internal rim of a young crater) areas, on the basis that most amateur images of the Moon are too dark.

Among the best references are Apollo-era images. This film photography shows the real contrast and brightness of the Moon. If we plan to render realistic images, at the price of the discernibility of lunar features, they are an invaluable source of inspiration. The images are widely available, thanks to NASA. The Project Apollo Archive is one of the most comprehensive sources:

www.apolloarchive.com/apollo_gallery.html

Lunar Reconnaissance Orbiter (LRO) images are mainly intended for scientific use, and numerous images are very contrasted and dark, since it was not intended that they would be directly printed. Another reliable inspiration is from NASA/ GSFC/Arizona State University with the famous, huge LRO WAC (Lunar Reconnaissance Orbiter Wide Angle Camera) image:

http://wms.lroc.asu.edu/lroc_browse/view/WAC_GL000

The contrast is not very strong (Figure 12.4), while the average surface is rather light. Note that pyroclastics are not so dark. Such features can be seriously darkened if they are the main subject of the image, as long as

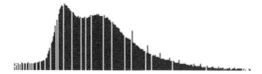

Figure 12.4 The histogram of the LRO WAC image shows a compressed contrast, with a great number of intermediate levels centered far away from the absolute black, and a great margin for high levels (which are not very frequent) to emphasize steep topographic features. This decreases the differentiation of terrains, but guarantees that the darkest levels will be preserved, even if the density is increased by offset printing.

neighboring highlands or maria remain bright enough (a third of the maximal brightness or so). Studying this image reveals that a large interval of high levels has been preserved to represent the most prominent topographic features (ray systems, craters at the poles, steep mountains). Perhaps this is not the best choice for an artistic rendering, but, obviously, the image's readability is perfect. Adopting the same chart of levels for an image to be published in a magazine is a very safe approach.

12.5 Preparing images for the Web

The images are mostly compressed in JPEG format to save storage capacity on servers and bandwidth while consulting the webpages. Other compressions (e.g. PNG, progressive JPEG, JPEG2000) are rarely used. Lunar images are rather large, especially mosaics. If we consider the very wide variety of browsing devices (smartphones, netbooks, desktops with large screens, assuming that the screens are uncalibrated), a good precaution is to standardize the images with a calibration chart (Section 7.1) and to limit the format, e.g. to 800 × 600, possibly with a link to load the image at a higher resolution. For instance, we have to consider that numerous netbooks have a 1366 × 768 resolution and the webpage contains a header, a footer, and one or several frames: there is little room left for the image. Reducing the resolution, possibly with the addition of a slight subsequent unsharp mask, may result in a smaller and sharper image.

A sidebar can include information about the feature, its location, technical information about the shooting, a simple calibration chart for the user to ensure that he or she is able to distinguish the whole range of levels,[6] the date, and the author's name – a private author may use the © symbol for copyright, but not the ® or ™ symbols, which refer to trademarks, registered or unregistered. The © symbol no longer has any legal status, but it informs the reader about the existence of an

---

[6] The chart may be placed on the front page of the website; see Figure 7.1.

author's rights.[7] A copyright notice may be added in the footer. The ownership of an astronomical image may be asserted by demonstrating the existence of the original, stacked image or movie. Large and centered watermarks are badly regarded!

12.6 Matching printing formats and sensor formats

Graphic artists often speak about a printing resolution, for instance the classic resolution of 300 DPI or dots per inch, while astrophotographers are interested in the number of pixels. We use different terms here: the *resolution* still refers to the accuracy of the telescope in arcseconds or the lunar image in arcseconds per pixel; the *resolution* for a graphic designer is the number of points contained in a printed line; while the *number of pixels* corresponds to the number of pixels of the image, not the number of photosites. This is an issue because printing in 300 DPI requires large numbers of pixels, as specified in Table 12.2.

We can see that only broadcast movie cameras, DSLRs, and some high-end sensors can produce images for large-format printing. For this purpose 300 DPI is a *de facto* standard because the image is accurate when it is viewed from a distance of about 25 cm (10 in), such as when we read a document.

If we reverse the reasoning, Table 12.3 indicates the possible printing formats for the usual numbers of pixels of planetary cameras. The output format with a 120% zoom is mentioned because it is commonly considered a reasonable value by professional graphic designers using classic enlargement algorithms.

Since amateur astronomers have better performing algorithms in their specialized software, an enlargement of 140%–150% is possible. As these algorithms are

Table 12.2 The theoretical number of pixels of an image to match the commonly desired "resolution" (DPI) for the most common ISO printing formats, assuming that the printer is able to print the full page with no margins (photo printers). This greatly exceeds the capability of most of the sensors in common use among amateurs, with the exception of some large sensors, e.g. the 3k IMX178, 4k CMOSIS, 6k Sony sensors, and, of course, DSLRs.

ISO printing format	Millimeters	Inches	Total number of pixels for 300 DPI
A2	594 × 420 mm	23.4 × 16.5 in	7020 × 4950
A3	420 × 297 mm	16.5 × 11.7 in	4950 × 3510
A4	297 × 210 mm	11.7 × 8.27 in	3510 × 2481
A5	210 × 148 mm	8.27 × 5.83 in	2481 × 1749

[7] There is no universal copyright agreement, and various countries have their own laws about the protection of creative works; the main treaties are the Buenos Aires Convention and the Berne Convention for the Protection of Literacy and Artistic Works.

Table 12.3 Printing format for publishing at 300 DPI and a reasonable 120% zoom for common sensors. In practice, amateurs having a megapixel camera (at least 1000 pixels in one dimension) tend to enlarge images up to A5 or A4. This means that the details are larger than the smallest points of the printed image.

Image number of pixels (or number of photosites for monochrome/ undebayered sensors)	Printing format at 300 DPI with 120% enlargement chart, viewed from a distance of 25 cm (values in centimeters)	Printing format at 300 DPI with 120% enlargement, viewed from a distance of 10 in (values in inches)
640 × 480	6.5 × 4.9	2.6 × 1.9
800 × 600	8.1 × 6.1	3.2 × 2.4
1280 × 1024	13.0 × 10.4	5.1 × 4.1
1936 × 1216	19.7 × 12.4	7.7 × 4.9
1616 × 1232	16.4 × 12.5	6.5 × 4.9
2048 × 1088	20.8 × 11.1	8.2 × 4.4
2048 × 2048	20.8 × 20.8	8.2 × 8.2

often developed for specific purposes, and because the Moon shows lines (rilles), quasi-pointlike features (peaks, craterlets), curves (craters), and irregular patterns (ruined craters, hills near Vallis Alpes), the algorithms turn out to provide almost identical results, with small, but appreciable, differences in noise. Here is a short list of algorithms provided by astronomical freeware; all of them perform well enough to resize an image up to 140%:

- Bicubic (all image editors)
- Lanczos
- Mitchell
- B-Spline, Spline
- Neuro
- NoHalo
- Dsinc

The only algorithm for enlargement without any alteration of the image is the linear resize. Pixel size is simply multiplied in width and height without creating interpolated (artificial) pixels.

The last thing to consider is the distance of vision to the print, and this has an enjoyable consequence. When we look at a large print, we tend to move our eye away from the image in order to have a comfortable field of view (Table 12.4). The smallest visible detail depends on the accuracy of our eye, and, of course, while the accuracy is constant, the smallest details fuse when we are too far distant, or, in other words, printed details have to be enlarged when we look at the print from a distance. That is why we can enlarge a printed image for an exhibition when it is close to A4 format or larger.

Table 12.4 ISO formats and possible enlargement for printing. The output formats are valid for display in an exhibition, not for publishing in a magazine.

ISO printing format	Comfortable distance for viewing	Reasonable enlargement factor/ printing DPI value	Minimum number of pixels of original image (may be a mosaic)
A5	25 cm/10 in	120%/250 DPI	2260 × 1470
A4	40 cm/16 in	140%/215 DPI	2540 × 1690
A3	60 cm/24 in	210%/140 DPI	2480 × 1650
A2	80 cm/32 in	280%/105 DPI	2480 × 1650

If we want to print in a very large format, possibly with a laser printer, which is not at all suited to delicate shades on large surfaces, a very interesting pre-processing in half tone must be carried out. Some image editors provide such functions (e.g. GIMP: Filters → Distorts → Newsprint); it is also available in on-line editors (Appendix 2).

12.7 Papers and inks

Laser printing hardly renders color/monochrome gradients on large surfaces, while home ink-jet photo printers and photo papers are intended to print vivid color images. Lunar eclipses, conjunctions, earthshines, halos, and petrographic images are perfectly rendered once the ink-jet printer has been calibrated with respect to the computer screen with the help of a chart (or a screen probe). Color inks with half shades guarantee a better rendition of images; that is why six-tone – or more – ink cartridges are preferable (e.g. yellow + two shades of magenta + two shades of cyan + black, possibly in addition to one or two shades of gray; sometimes a transparent, protective-layer cartridge is used on top of the colors). In these cases, glossy paper is well adapted and the color balance shows little alteration with time, as long as the prints are not exposed to direct sunlight. Nonetheless, an inevitable, differen-tial alteration of ink components can lead to a greenish image after some months or years, even if the image is sheltered.

The best way, which is more expensive but worth the effort, is professional printing with a digital minilab using classic Kodak RA-4 chemical photographic printing. Professional, Ilfochrome/Cibachrome printing is the most durable (and also expensive) process. Chemical printing is feasible for a photographer – this is the most reliable solution because the printing of astronomical images is rather specialized – or it can be done via the Web. The issue with Web printing is that laboratories generally increase the density (darkening) to obtain a better satura-tion, resulting in more vivid colors. Unfortunately, this rarely matches the needs of astronomical images, which often contain delicate shades, especially at low levels. Another solution is using professional, durable inks like K3/K4 UltraChrome, ChromaLife 100, or comparable products. That solution is rather expensive. Moreover, the nozzles of consumer ink-jet printers tend to clog, especially in summer, unless the printer is regularly in use. This is why it is better to choose durable, color printing by professionals, preferentially at the photographer's.

Black-and-white ink-jet printing uses a mixture of black and other colors, once again leading to a slow but inevitable alteration: just like color images, the black-and-white image often tends to become greenish after some months, whatever the paper (glossy, matte, standard). It is hugely preferable to print in pure black-and-white mode if the printer driver allows it. Ink-jet printer drivers forbid pure black-and-white printing of glossy paper because the latter requires the application of a protective layer in normal, color printing. Using a glossy paper while selecting a matte/ordinary paper while printing in pure black-and-white is possible, but the lack of a protective layer results in an extremely fragile print: a fingernail easily scratches the image. Fortunately, an excellent, economical, and very durable solution is printing on satin/luster photo paper (not ordinary matte paper) in pure, black-and-white mode: prints remain unaltered for years. Of course, since the print is rather matte, it seems less contrasted, but it is durable and perfect for an exhibition because reflections of light never happen: the prints can be comfortably observed at all angles, even under inappropriate lighting. Another solution is using monochrome inks like K7 (for this one needs the QuadTone RIP shareware, or one can consult piezography K7/K6 ink dealers), which are compatible with matte, glossy, and baryta papers, giving superb results. The inks temporarily replace the color inks in compatible printers.

Appendix 1

Maps of the Moon, the Lunar 100, and other targets

Map of the Moon

Figure A.1 shows a simple map with Rükl's map grid (1–76), simplified seleno-graphic coordinates (in degrees), and the Latin names given by the International Astronomical Union.

Lunar 100

The Lunar 100 (Table A.1) lists a hundred features to observe on the Moon at various scales. This is equivalent to the Messier Catalog for deep-sky observers and was established in the prominent article by Charles A. Wood published in *Sky and Telescope* magazine, April 2004 (www.skyandtelescope.com/observing/the-lunar-100). This list includes various targets sorted by difficulty, ranging from large features to accurate details. Intended primarily for observers, it can also be used for imaging challenges, or simply on account of its educational qualities. The whole tour may last several years because of the increasing difficulty of observation or identification due to small sizes and poor contrast, extreme polar locations, and librations.

Figure A.1 Map of the Moon.

Table A.1 The Lunar 100 by Charles A. Wood (from Wikipedia)

L	Feature name	Significance	Latitude	Longitude	Diameter (km)
L1	Moon	Large satellite	—	—	3476
L2	Earthshine	Twice-reflected sunlight	—	—	—
L3	Mare/highland dichotomy	Two materials with distinct compositions	—	—	—
L4	Apennines	Imbrium basin rim	18.9° N	3.7° W	70
L5	Copernicus	Archetypal large complex crater	9.7° N	20.1° W	93
L6	Tycho	Large rayed crater with impact melts	43.4° S	11.1° W	85
L7	Altai Scarp	Nectaris basin rim	24.3° S	22.6° E	425
L8	Theophilus, Cyrillus, Catharina	Crater sequence illustrating stages of degradation	13.2° S	24.0° E	—
L9	Clavius	Lacks basin features in spite of its size	58.8° S	14.1° W	225
L10	Mare Crisium	Mare contained in large circular basin	18.0° N	59.0° E	540
L11	Aristarchus	Very bright crater with dark bands on its walls	23.7° N	47.4° W	40
L12	Proclus	Oblique-impact rays	16.1° N	46.8° E	28
L13	Gassendi	Floor-fractured crater	17.6° S	40.1° W	101
L14	Sinus Iridum	Very large crater with missing rim	45.0° N	32.0° W	260
L15	Straight Wall	Best example of a lunar fault	21.8° S	7.8° W	110
L16	Petavius	Crater with domed and fractured floor	25.1° S	60.4° E	177
L17	Schröter's Valley	Giant sinuous rille	26.2° N	50.8° W	168
L18	Mare Serenitatis dark edges	Distinct mare areas with different compositions	17.8° N	23.0° E	—
L19	Alpine Valley	Lunar graben	49.0° N	3.0° E	165
L20	Posidonius	Floor-fractured crater	31.8° N	29.9° E	95
L21	Fracastorius	Crater with subsided and fractured floor	21.5° S	33.2° E	124
L22	Aristarchus Plateau	Mysterious uplifted region mantled with pyroclastics	26.0° N	51.0° W	150
L23	Pico	Isolated Imbrium basin-ring fragment	45.7° N	8.9° W	25
L24	Hyginus Rille	Rille containing rimless collapse pits	7.4° N	7.8° E	220
L25	Messier and Messier A	Oblique ricochet-impact pair	1.9° S	47.6° E	11
L26	Mare Frigoris	Arcuate mare of uncertain origin	56.0° N	1.4° E	1600
L27	Archimedes	Large crater lacking central peak	29.7° N	4.0° W	83
L28	Hipparchus	First drawing of a single crater	5.5° S	4.8° E	150
L29	Ariadaeus Rille	Long, linear graben	6.4° N	14.0° E	250
L30	Schiller	Possible oblique impact	51.9° S	39.0° W	180
L31	Taruntius	Young floor-fractured crater	5.6° N	46.5° E	56
L32	Arago Alpha and Beta	Volcanic domes	6.2° N	21.4° E	26

Table A.1 (cont.)

L	Feature name	Significance	Latitude	Longitude	Diameter (km)
L33	Serpentine Ridge	Basin inner-ring segment	27.3° N	25.3° E	155
L34	Lacus Mortis	Strange crater with rille and ridge	45.0° N	27.2° E	152
L35	Triesnecker Rilles	Rille family	4.3° N	4.6° E	215
L36	Grimaldi basin	A small two-ring basin	5.5° S	68.3° W	440
L37	Bailly	Barely discernible basin	66.5° S	69.1° W	303
L38	Sabine and Ritter	Possible twin impacts	1.7° N	19.7° E	30
L39	Schickard	Crater floor with Orientale basin ejecta stripe	44.3° S	55.3° W	227
L40	Janssen Rille	Rare example of a highland rille	45.4° S	39.3° E	190
L41	Bessel ray	Ray of uncertain origin near Bessel	21.8° N	17.9° E	—
L42	Marius Hills	Complex of volcanic domes and hills	12.5° N	54.0° W	125
L43	Wargentin	A crater filled to the rim with lava or ejecta	49.6° S	60.2° W	84
L44	Mersenius	Domed floor cut by secondary craters	21.5° S	49.2° W	84
L45	Maurolycus	Region of saturation cratering	42.0° S	14.0° E	114
L46	Regiomontanus central peak	Possible volcanic peak	28.0° S	0.6° W	124
L47	Alphonsus dark spots	Dark-halo eruptions on crater floor	13.7° S	3.2° W	119
L48	Cauchy region	Fault, rilles, and domes	10.5° N	38.0° E	130
L49	Gruithuisen Delta and Gamma	Volcanic domes formed with viscous lavas	36.3° N	40.0° W	20
L50	Cayley Plains	Light, smooth plains of uncertain origin	4.0° N	15.1° E	14
L51	Davy crater chain	Result of comet-fragment impacts	11.1° S	6.6° W	50
L52	Crüger	Possible volcanic caldera	16.7° S	66.8° W	45
L53	Lamont	Possible buried basin	4.4° N	23.7° E	106
L54	Hippalus Rilles	Rilles concentric to Humorum basin	24.5° S	2 9.0° W	240
L55	Baco	Unusually smooth crater floor and surrounding plains	51.0° S	19.1° E	69
L56	Australe basin	A partially flooded ancient basin	49.8° S	84.5° E	880
L57	Reiner Gamma	Conspicuous swirl and magnetic anomaly	7.7° N	59.2° W	70
L58	Rheita Valley	Basin secondary-crater chain	42.5° S	51.5° E	445
L59	Schiller–Zucchius basin	Badly degraded overlooked basin	56.0° S	45.0° W	335
L60	Kies Pi	Volcanic dome	26.9° S	24.2° W	45
L61	Mösting A	Simple crater close to center of lunar near side	3.2° S	5.2° W	13
L62	Rümker	Large volcanic dome	40.8° N	58.1° W	70

Table A.1 (cont.)

L	Feature name	Significance	Latitude	Longitude	Diameter (km)
L63	Imbrium sculpture	Basin ejecta near and overlying Boscovich and Julius Caesar	11.0° N	12.0° E	—
L64	Descartes	Apollo 16 landing site; putative region of highland volcanism	11.7° S	15.7° E	48
L65	Hortensius domes	Dome field north of Hortensius	7.6° N	27.9° W	10
L66	Hadley Rille	Lava channel near Apollo 15 landing site	25.0° N	3.0° E	—
L67	Fra Mauro formation	Apollo 14 landing site on Imbrium ejecta	3.6° S	17.5° W	—
L68	Flamsteed P	Proposed young volcanic crater and Surveyor 1 landing site	3.0° S	44.0° W	112
L69	Copernicus secondary craters	Rays and craterlets near Pytheas	19.6° N	19.1° W	4
L70	Humboldtianum basin	Multi-ring impact basin	57.0° N	80.0° E	650
L71	Sulpicius Gallus dark mantle	Ash eruptions northwest of crater	19.6° N	11.6° E	12
L72	Atlas dark-halo craters	Explosive volcanic pits on the floor of Atlas	46.7° N	44.4° E	87
L73	Smythii basin	Difficult-to-observe basin scarp and mare	2.0° S	87.0° E	740
L74	Copernicus H	Dark-halo impact crater	6.9° N	18.3° W	5
L75	Ptolemaeus B	Saucer-like depression on the floor of Ptolemaeus	8.0° S	0.8° W	16
L76	W. Bond	Large crater degraded by Imbrium ejecta	65.3° N	3.7° E	158
L77	Sirsalis Rille	Procellarum basin radial rilles	15.7° S	61.7° W	425
L78	Lambert R	A buried "ghost" crater	23.8° N	20.6° W	54
L79	Sinus Aestuum	Eastern dark-mantle volcanic deposit	12.0° N	3.5° W	90
L80	Orientale basin	Youngest large impact basin	19.0° S	95.0° W	930
L81	Hesiodus A	Concentric crater	30.1° S	17.0° W	15
L82	Linné	Small crater once thought to have disappeared	27.7° N	11.8° E	2.4
L83	Plato craterlets	Crater pits at limits of detection	51.6° N	9.4° W	101
L84	Pitatus	Crater with concentric rilles	29.8° S	13.5° W	97
L85	Langrenus rays	Aged ray system	8.9° S	60.9° E	132
L86	Prinz Rilles	Rille system near the crater Prinz	27.0° N	43.0° W	46
L87	Humboldt	Crater with central peaks and dark spots	27.0° S	80.9° E	207

Table A.1 (cont.)

L	Feature name	Significance	Latitude	Longitude	Diameter (km)
L88	Peary	Difficult-to-observe polar crater	88.6° N	33.0° E	74
L89	Valentine Dome	Volcanic dome	30.5° N	10.1° E	30
L90	Armstrong, Aldrin, and Collins	Small craters near the Apollo 11 landing site	1.3° N	23.7° E	3
L91	De Gasparis Rilles	Area with many rilles	25.9° S	50.7° W	30
L92	Gyldén Valley	Part of the Imbrium radial sculpture	5.1° S	0.7° E	47
L93	Dionysius rays	Unusual and rare dark rays	2.8° N	17.3° E	18
L94	Drygalski	Large South-pole-region crater	79.3° S	84.9° W	162
L95	Procellarum basin	The Moon's biggest basin?	23.0° N	15.0° W	3200
L96	Leibnitz Mountains	Rim of South Pole–Aitken basin	85.0° S	30.0° E	—
L97	Inghirami Valley	Orientale basin ejecta	44.0° S	73.0° W	140
L98	Imbrium lava flows	Mare lava-flow boundaries	32.8° N	22.0° W	—
L99	Ina	D-shaped young volcanic caldera	18.6° N	5.3° E	3
L100	Mare Marginis swirls	Possible magnetic deposits	18.5° N	88.0° E	—

Lunar-mission landing sites

Human-made objects on the Moon are about 250 times smaller than the best resolution we can achieve with an amateur telescope.[1] Nonetheless, the hunt for the locations of the landing sites (Table A.2 and Figure A.2) is interesting because of the variety of those areas, and this is a lesson both in selenology and in history. Some sites are mentioned in the Lunar 100 list. A complete list of crashes and landing sites is given in Appendix 2. Another way to search for the sites is by using the Virtual Moon Atlas. The menu File → Database opens a list. We can look for rows labeled "HIS" in the leftmost column (or use Edit → selection, then choose just "Probe," "Human mission," and "Inert equipment"). A click on the row centers the lunar map of the main window to the selected site. Once the site has been accurately located on the map and on our image, the best thing to do is to indicate the exact location with a cross having a clear center, or two separated lines at right angles, so as not to mask out the site.

---

[1] The Lunar Reconnaissance Orbiter (LRO) distinctly imaged a number of probes and equipment left in place by Apollo crews. It even permitted the visual recovery of the formerly lost Lunokhod 1. The LRO's two narrow-field cameras have a resolution of 0.5 m on the Moon. An amateur's image may sometimes reach a resolution of 400 m (about 0.2 arcsecond) or a bit better, as of 2016. Viewed from the Hubble Space Telescope, a 4-m object on the Moon subtends an apparent angle of 0.002 arcsecond.

Table A.2 Some landing sites and the site of first intentional impact. Since equatorial orbits consume less propellant and sites having the Earth in the line of sight have no obstacles to radio communication, numerous sites are located not far from the equator. The list is sorted to help with the planning of imaging sessions with respect to the colongitude. Some impacts are not listed because they are not in the line of sight, such as those by LCROSS/ Centaur (324.5° E, 84.9° S) and Moon Impact Probe/Chandrayaan 1 (30° W, 89° S).

Mission	Longitude	Latitude	Notes
Luna 24	62.20° E	12.25° N	Lander and sample return
Luna 20	56.50° E	3.57° N	Lander and sample return
Luna 16	56.30° E	0.68° S	Lander and sample return
Apollo 17	30.77° E	20.19° N	Manned landing
Luna 21 + Lunokhod 2	30.38° E	25.51° N	Lander and rover
Apollo 11	23.47° E	0.67° N	First manned landing, 1969
Surveyor 5	23.18° E	1.41° N	Lander
Apollo 16	15.49° E	8.97° S	Manned landing
Apollo 15	3.63° E	26.13° N	Manned landing
Luna 2	0.00°	29.10° N	First impact, 1959
Surveyor 6	1.37° W	0.46° N	Lander
Surveyor 7	11.41° W	41.01° S	Lander
Apollo 14	17.47° W	3.64° S	Manned landing
Chang'e 3	19.51° W	44.12° N	Lander and rover
Surveyor 3	23.34° W	2.94° S	Lander
Apollo 12	23.42° W	3.01° S	Manned landing
Luna 17 + Lunokhod 1	35.00° W	38.28° N	Lander and rover
Surveyor 1	43.21° W	2.45° S	Lander
Luna 13	62.05° W	18.87° N	Lander
Luna 9	64.37° W	7.08° N	First landing, 1966

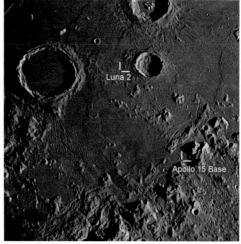

Figure A.2 Some landing sites' locations. They were located with the help of the Virtual Moon Atlas.

Appendix 2

Webpages, books, and freeware for the Moon

Professional lunar images

Professional images show features at various angles and resolutions. This greatly helps to interpret amateur photographs, especially to detect artifacts from over-processing. They are also a source of inspiration to make amateur images resemble the real Moon taken by orbital manned and robotic missions, with genuine contrasts.

www.lroc.asu.edu	A superb site by NASA and Arizona State University about the Moon, especially the Lunar Reconnaissance Orbiter (LRO) camera.
http://lunar.gsfc.nasa.gov	A site by NASA and Goddard Space Flight Center about the LRO. Also provides useful links for the Moon and planetology.
http://nssdc.gsfc.nasa.gov/planetary/planets/moonpage.html	Lunar missions from various space agencies
www.lpi.usra.edu/resources/lunarorbiter/	The Lunar and Planetary Institute hosts a magnificent collection of LRO images
www.kaguya.jaxa.jp/index_e.htm	JAXA's Kaguya (SELENE) and its beautiful images
www.isro.gov.in/Spacecraft/chandrayaan-1	ISRO's Chandrayaan-1
http://ser.sese.asu.edu	The Space Exploration Resources site by Arizona State University includes numerous images and data from unmanned and manned space missions.
www.nasa.gov/mission_pages/apollo/revisited/index.html#.VT47B5M2JxA	Close-up of Apollo landing sites by NASA's LRO
http://en.wikipedia.org/wiki/List_of_artificial_objects_on_the_Moon	A well-documented list of artificial objects on the Moon since 1959, with coordinates
www.mapaplanet.org/explorer/moon.html	Map-a-Planet (including the Moon) by the USGS, multispectral data and maps
http://planetarynames.wr.usgs.gov/Page/Moon1to1MAtlas	
http://astrogeology.usgs.gov/maps	
www.moonzoo.org	The Moon viewed by the LRO
www.google.com/moon	Visit the Moon
www.lpi.usra.edu/resources/mapcatalog/usgs	Geologic Atlas of the Moon
http://target.lroc.asu.edu/q3	NASA's Act-React QMAP

Some amateur lunar images

Countless lunar images are shared by talented amateur astrophotographers. This short list is only a sample to illustrate the variety of their inspiration.

www.makolkin.ru/Gallery/gallery.html	Dmitry Makolkin (Russia)
www.damianpeach.com	Damian Peach (UK)
www.astrophoto.fr/index.html	Thierry Legault (France)
http://higginsandsons.com/astro	Wes Higgins (USA)
http://danikxt.blogspot.fr/p/luna.html	Dani Caxete (Spain)
www.astronominsk.org/index_en.htm	Yuri Goryachko, Mikhail Abgarian, Konstantin Morozov (Belarus)
www.visit-the-moon.com	Andrew M. Bray (Australia)
www.lunar-captures.com	George Tarsoudis (Greece)
www.lazzarotti-hires.com	Paolo Lazzaroti (Italy)
www.astrosurf.com/viladrich	Christian Viladrich (France)
http://moonscience.yolasite.com/	Maurice Collins (New Zealand)
www.digitalsky.org.uk	Pete Lawrence (England)
http://sfire.astroclub.kiev.ua/tag/Moon	Pavel Presnyakov (Ukraine)
www.facebook.com/robert.reeves.7773	Robert Reeves

Some personal pages about industrial/planetary cameras

http://damien.douxchamps.net/ieee1394/cameras	A very accurate list of Firewire cameras by Damien Douxchamps
http://nicolas.dupontbloch.free.fr/camera.htm	A sample list of affordable planetary cameras
www.rkblog.rk.edu.pl/astro/kamery-ccd	A sample list of astronomical cameras along with a blog

Organizations for lunar imaging and observation

Renowned associations and organizations encourage the observation and shooting of the Moon (and planets). They provide abundant documentation and periodic reports.

http://observethemoonnight.org	International Observe the Moon Night (InOMN), in association with NASA, is a worldwide, annual event to encourage the observation and understanding of the

(cont.)

	Moon. Associations can register their event. Information and educational material are provided.
http://alpo-astronomy.org/	The Association of Lunar and Planetary Observers (the Japanese ALPO seems to be devoted to planets only)
www.baalunarsection.org.uk/tnm.htm	The Lunar Section of the British Astronomical Association
www.amlunsoc.org	The American Lunar Society
http://occultations.org/	The International Occultation Timing Association
www.euraster.net	Asteroidal Occultations in Europe; also refers to lunar occultations.

Select bibliography

Books about the Moon:

- *Apollo over the Moon, a View from Orbit*, Harold Masursky, G. W. Colton, and Farouk El-Baz (http://history.nasa.gov/SP-362)
- *Atlas of the Lunar Terminator*, John E. Westfall (Cambridge University Press)
- *Atlas photographique de la Lune* (in French), Jérôme Grenier (on-line at http://jeromegrenier.free.fr/atlas.htm)
- *Kaguya Lunar Atlas*, Motomaro Shirao and Charles A. Wood (Springer), close-ups from the Kaguya/Selene lunar probe.
- Lunar Picture Of the Day – no longer maintained but still a fantastic tour of the Moon by Chuck Wood (http://www2.lpod.org/)
- *Moon Atlas*, Antonín Rükl, with additions, re-published (Sky Publishing Corp.)
- On-line, multi-contributor encyclopedia: http://the-moon.wikispaces.com/Introduction
- *Photographic Atlas of the Moon*, Siew Meng Chong, Albert C. H. Lim, and Poon Seng Ang (Cambridge University Press)
- *Photographic Moon Book*, Alan Chu (www.alanchuhk.com)
- *The Cambridge Photographic Moon Atlas*, Alan Chu, Wolfgang Paech, and Mario Weigand (Cambridge University Press)
- *The Clementine Atlas of the Moon*, Ben Bussey and Paul D. Spudis (Cambridge University Press)
- *The Geologic History of the Moon*, Don E. Wilhelms, USGS (on-line at USGS)
- *The Modern Moon*, Charles A. Wood (Sky Publishing Corp.)
- *The Moon and How to Observe It*, Peter Grego (Springer)
- *The Moon in Close-up*, John Wilkinson (Springer)

Some recent books about general astrophotography containing information about lunar imaging:

- *A Guide to DSLR Planetary Imaging on CD-ROM*, Jerry Lodriguss (www.astropix .com/GDPI/GDPI.htm)
- *Astrophotography*, Thierry Legault (Rocky Nook), includes 22 pages (in the 2014 edition) about Moon and planetary imaging, planetary cameras, and planetary telescopes
- *Practical Astrophotography*, Jeffrey R. Charles (Springer), includes 5 pages about imaging lunar eclipses (in the 2013 edition)
- *Scientific Astrophotography*, Gerald Hubbel (Springer), includes 11 pages about specific lunar imaging processing and tens of pages about equipment including lunar imaging (in the 2013 edition)
- *The Backyard Astronomer's Guide*, Terence Dickinson and Alan Dyer (Firefly books), includes a chapter (in the 2010 edition) about imaging the Moon and Sun
- *The Handbook of Astronomical Image Processing*, Richard Berry and James Burnell (Willmann-Bell), includes 15 pages (in the 2nd edition) about imaging the Moon, planets, and Sun

Processing and image-acquisition freeware for lunar images

This short list contains useful freeware adapted to lunar image processing and ephemeris and image acquisition with various operating systems.

www.autostakkert.com	AutoStakkert! 2 is renowned for lunar stacking, by Emil Kraaikamp
www.avistack.de	AviStack perfectly stacks lunar images, by Michael Theusner and Joe Zawodny
www.obs-psr.com/astrolinux	Bootable Linux for astronomy (CD or USB key, no installation needed)
www.fitswork.de/software	Fitswork is a deep-sky image editor, some functions may be useful for lunar imaging, by J. Dierks
www.gimp.org	GIMP is a powerful image editor. Version 2.8 manages 8-bit images only, but a 16-bit-image version is planned.
www.myastroshop.com.au/guides/gstar	Gstar-EX video camera and its acquisition software, for surveying transient lunar phenomena

(cont.)

www.astrosurf.com/happix	Happix is no longer maintained, but it is free, light, and devoted to astronomy, by Vincent Cotrez
http://research.microsoft.com/en-us/downloads/69699e5a-5c91-4b01-898c-ef012cbb07f7	Image Composite Editor helps to create panoramic images and mosaics, by Microsoft
http://imagej.nih.gov/ij/index.html	ImageJ is a multi-platform, Java-based image editor for astronomy
http://jaggedplanet.com/imerge.html	iMerge perfectly merges images for lunar mosaics, by Jon Grove
www.astrosurf.com/buil/us/iris/iris.htm	IRIS is one of the best astronomical image-processing software packages (it is a bit complicated, but worth the effort), by Christian Buil
www.distroastro.org	Linux for astronomy
https://ltvt.wikispaces.com/LTVT	Lunar Terminator Visualization Tool
http://sourceforge.net/directory/science-engineering/astronomy/os:linux/freshness:recently-updated/	Numerous excellent items of astronomy software for Linux
http://photofiltre.free.fr	PhotoFiltre is a simple-to-use but very good image editor, by Antonio Da Cruz
www.astronomie.be/registax	RegiStax performs stacking and remarkable sharpness processing, by Cor Berrevoets
www.stellarium.org/fr	Stellarium is a brilliant multi-platform planetarium
http://traitement-d-images-tim.webnode.fr	Tim is an unusual and user-friendly image editor
http://ap-i.net/avl/fr/start	Virtual Moon Atlas: the reference in lunar cartography software, by Patrick Chevalley and Christian Legrand
www.clearskyinstitute.com/xephem	XEphem Ephemerides, for Unix, by Elwood Downey
http://firecapture.wonderplanets.de/	FireCapture is an acquisition software compatible with numerous astronomical and industrial cameras, by Torsten Edelmann
www.emmanuel-rietsch.fr/Astronomie/index_EN.htm	Video Sky 2011 is image-acquisition software for astronomical and industrial cameras, by Emmanuel Rietsch

Other technical and specific graphical freeware

RAMDisk utilities may improve performances for image acquisition:

www.tekrevue.com/tip/how-to-create-a-4gbs-ram-disk-in-mac-os-x	How to create a RAMDisk on Mac OS X, by J. Tanous
https://bogner.sh/projects	Direct link to F. Bogner's RAMDisk creator for Mac OS X
www.tekrevue.com/tip/create-10-gbs-ram-disk-windows	The same trick for Windows, by J. Tanous
www.raymond.cc/blog/12-ram-disk-software-benchmarked-for-fastest-read-and-write-speed/	A benchmark and links to RAMdisk software, by Raymond.CC

Here are some utilities to manage image files and movie files. Some of the freeware developed by professionals relies on acceptance of third-party software, which is not always stated explicitly.

www.faststone.org/index.htm	FastStone is an image browser with editing and conversion functionalities, by FastStone Soft
www.koyotesoft.com	Free HD Converter converts AVCHD movies to common video formats and Free Video Converter converts various video formats, by Koyote Soft
http://starrydave.com/	Dave's utilities for solar imaging may be helpful for lunar eclipses like "Dave's Video Stabilizer," by Starry Dave
https://sites.google.com/site/astropipp/	PIPP Image preprocessor, including video stabilization and debayering, by Chris Garry
www.videotovideo.org	Video-to-video converter by Media Converters
www.videolan.org	VLC media player reads numerous video formats, including AVCHD, by VideoLAN organization
www.xnview.com/en	XnView is a multi-platform image browser with editing and conversion functionalities, including FITS format, by XnSoft
http://deepskystacker.free.fr/english/index.html	DeepSkyStacker Live helps to focus on a computer screen when the DSLR has no LiveView functionality, by Luc Coiffier

Appendix 3
Figure data

The main technical data regarding image acquisition in Table A.3 are given only when appropriate. Note that most images have been severely cropped to ensure the legibility of the central subject at the expense of the original resolution.

Table A.3 Technical data of images

Figure	Optics	Camera	Accessories
1.1	200-mm Celestron 8	Nikon FE, Ilford FP4 film	Celestron focal reducer/field corrector F/6.3
1.2	Synta 254-mm (10-in) Newton	QHYCCD QHY5	Celestron Ultima Barlow, Astronomik type II-C red filter
1.5	Vixen 70/420 Fraunhofer refractor	Smartphone held by hand	Celestron Ultima 8-mm eyepiece
1.9	Meade ETX90 Mak	ZWO ASI120MM	
1.11	1000-mm (40-in) DIY motorized Dobsonian	The Imaging Source DMK31 monochrome	Baader 742-nm infrared-pass filter, 3× Barlow lens Image Frédéric Géa
1.14	Synta 254-mm (10-in) Newtonian	QHYCCD QHY5	Barlow lens, red filter
2.1	Synta 254-mm (10-in) Newtonian	ZWO ASI120MM	Barlow lens, red filter Test image
2.2	Synta 254-mm (10-in) Newtonian	ZWO ASI120MM	Part of Figure 3.16
2.3 top	Celestron 5 (125 mm/5 in)	I-Nova PLB-Cx	Televue 2× Barlow lens Image Etienne Martin
2.3 bottom	Synta 254-mm (10-in) Newtonian	ZWO ASI120MM	Barlow lens, red filter
2.8	Angénieux 180-mm telelens	ZWO ASI120MM	Red filter
2.11, 2.12	460-mm (18-in) Dobsonian, undriven	ZWO ASI120MM	Barlow lens, red filter
2.17 left	Lens covered	Aptina MT9M034 sensor	

Table A.3 (cont.)

Figure	Optics	Camera	Accessories
2.17 right	Test image, strongly zoomed in	Aptina MT9M001 sensor	
2.18	Built-in zoom	Canon HG10 HD camcorder	Test image, cropped
2.19	Celestron 5	Canon EOS 350D DSLR	Test image, cropped, Exaggerated contrast
2.24	Celestron 8	Philips ToUCam Pro webcam in optimized color mode	Test image with electromagnetic disturbances Barlow lens, red filter
2.26	Built-in 10× zoom	Canon HG10 HD camcorder	Test images
2.27	Built-in 20× zoom	JVC Everio GZ-HM300SE camcorder	Test images
2.28	Celestron 5	ToUCam Pro II	Test image at prime focus
2.29	Synta 254-mm/10-in Newton	ToUCam Pro II green layer only	Celestron Ultima 12-mm eyepiece
2.34	Celestron 5	Miniature video camera	Camera lens removed
2.35	18–105 Nikkor	Nikon D3200	
2.36	18–55 EFS Canon	Canon EOS 350D	
2.37	18–105 Nikkor	Nikon D3200	
2.38	Celestron 5	Canon EOS 350D	Mirror raised prior to shooting
3.2	Officina Stellare Hiper Apo 150 mm (6 in) F/8	Canon HG10 camcorder (digiscopy)	Test image TMP planetary II 8-mm eyepiece
3.4	Celestron 5	Samsung gt s5830 smartphone (digiscopy)	12.5-mm Ultima Celestron eyepiece Image by Mathias Barbarroux
3.13	Synta 254-mm/10-in Newtonian (test image)	ZWO ASI224MC	Barlow lens, red filter
3.15 left	Synta 254-mm/10-in Newtonian (test image)	QHYCCD QHY5	Barlow lens, red filter
3.15 right	Synta 254-mm/10-in Newtonian (test image)	ZWO ASI120MM	Barlow lens, red filter

Table A.3 (cont.)

Figure	Optics	Camera	Accessories
3.16	Synta 254-mm/10-in Newtonian (test image)	QHYCCD QHY5	Prime focus, infrared- and ultraviolet-rejection filter; 200 × 2.5-s frames
3.19	Officina Stellare Hiper Apo 150 mm (6 in) F/8	Canon EOS350D	Test image
3.21	Synta 254-mm/10-in Newtonian	ZWO ASI120MM	Test image Barlow lens, red filter
4.1	Synta 150-mm (6-in) achromat	Canon EOS 350D	Prime focus, 400 ISO, 0.1 s
4.12, 4.16, 4.22	Synta 254-mm/10-in Newtonian	ZWO ASI120MM	Cropped test images Barlow lens, red filter
4.24	Synta 254-mm/10-in Newtonian	QHYCCD QHY5	Test image Barlow lens, red filter
4.25	Synta 254-mm/10-in Newtonian	ZWO ASI120MM	Test image Barlow, red filter
5.1 left, center	18–105-mm Nikkor	Nikon D3200	
5.1 right, 5.2, 5.3, 5.4, 5.5, 5.6	18–55-mm EFS Canon	Canon EOS 350D	
5.8	Unrecorded photolens	Canon 5D or 7D	Image by Gilles Boutin
5.9	Angénieux 180 mm	Nikon D70	
5.10	Angénieux 180 mm	ZWO ASI120MM	DIY C/T2/Nikon adapter, green filter
5.12	70–300-mm Nikkor	Nikon D7000	Image by Dani Caxete
5.13	Celestron 5	Canon EOS 350D	
5.14	Synta 150-mm (6-in) achromat	Canon EOS 350D	
5.15	18–105-mm Nikkor	Nikon D3200	
5.16	Built-in zoom	Canon HG 10 HD camcorder	
5.17	18–55-mm EFS Canon	Canon EOS 350D	
5.18	Synta 150-mm (6-in) achromat	Canon EOS 350D	Attenuated saturation
5.19	7 × 50 finder, eyepiece was unscrewed	Philips Vesta Pro webcam	Modified driver for 1-s exposures
5.20	18–55-mm EFS Canon	Canon EOS 350D	Motorized mount
5.22, 5.23, 5.25	Celestron 5	Canon EOS 350D	5.22 unmotorized, others motorized
5.24	Vixen 70/420 Fraunhofer refractor	Canon EOS 350D	HDR, 800 ISO. 1: 1/4000 s;

Table A.3 (cont.)

Figure	Optics	Camera	Accessories
			2 and 3: 1/4000 s + 1/1000 s + 1/250 s + 1/100 s + 1/50 s + 1/10 s; 4 and 9: 1/250 s + 1/10 s + 1 s; 5 and 7: 1/10 s + 1 s + 10 s; 6: 1 s + 10 s + 20 s; 8: 1/10 s + 1 s; 10: 1/1000 s + 1/250 s + 1/10 s + 1 s; 11: 1/1000 s + 1/100 s + 1/5 s; 12: 1/4000 s + 1/1000 s + 1/250 s + 1/50 s + 1/10 s.
5.26, 5.27	Celestron 5	Cooled ZWO ASI120MM	Focal reducer/field corrector F/6.3
6.1	Synta 254-mm/10-in Newton	ZWO ASI120MM	Cropped test images Various magnifications, red filter
6.3	Celestron 5	Nikon D3200	Focal reducer/field corrector F/6.3
7.5	Synta 254-mm/10-in Newtonian	QHYCCD QHY5	Test images Barlow lens, red filter
7.6 to7.9, 7.11 to7.12	Synta 254-mm/10-in Newtonian	ZWO ASI120MM	Cropped images Barlow lens, red filter
7.13, 8.1	Synta 254-mm/10-in Newtonian	QHYCCD QHY5	Cropped images Barlow lens, red filter
8.2 to 8.10	Synta 254-mm/10-in Newtonian (test images)	ZWO ASI120MM	Barlow lens, red filter
8.11	Synta 254-mm/10-in Newtonian	ZWO ASI120MM	Cropped Barlow lens, red filter
8.12	Celestron 5	Canon EOS 350D	
8.13, 8.14 right	Synta 254-mm/10-in Newtonian	ZWO ASI120MM	Barlow lens, Astronomik Type II-c CCD imaging RGB filters
8-16	Celestron 5	Canon EOS 350D	
8.17, 8.18	Synta 254-mm/10-in Newtonian	ZWO ASI120MM	Barlow lens, red filter Mosaic

Table A.3 (cont.)

Figure	Optics	Camera	Accessories
8.19	Officina Stellare Hiper Apo 150 mm (6 in) F/8	QHYCCD QHY5	Infrared- and ultra-violet-rejection filter Mosaic
8.20, 8.21	Synta 254-mm/10-in Newtonian	ZWO ASI120MM	Barlow lens, red filter
9.2	230-mm (9-in) Santel	Unibrain Fire-i 702	Cropped Image by Yuri Goryachko, Mikhail Abgarian, and Konstantin Morozov
9.3, 9.12	150-mm/6-in Hiper Apo	QHYCCD QHY5	Infrared- and ultra-violet-rejection filter Mosaic
9.4, 9.5, 9.6 left, 9.7, 9.9 to 9.11	Synta 254-mm/10-in Newtonian	ZWO ASI120MM	Barlow lens, red filter Mosaic 9.7 series by Mathias Barbarroux
10.2 left, 10.3	Synta 254-mm/10-in Newtonian	ZWO ASI120MM	Barlow lens, red filter
11.1 to11.3, 11.6 to 11.10, 11.12, 11.16, 11.17 left, 11.19, 11.20 to 11.22, 11.26, 11.27	Synta 254-mm/10-in Newtonian	ZWO ASI120MM	Cropped Barlow lens, red filter 11.19 is a mosaic
11.18	Synta 254-mm/10-in Newtonian	ZWO ASI178MM	Cropped Barlow lens, red filter
11.4, 11.5, 11.11, 11.13 to 11.15, 11.17 right, 11.23 to 11.25, 11.28, 11.29	Synta 254-mm/10-in Newtonian	QHYCCD QHY5	Cropped Barlow lens, red filter
11.14	355-mm (14-in) Schmidt–Cassegrain	Video camera	
A.2 left, right	Synta 254-mm/10-in Newtonian	ZWO AI120MM	Cropped Barlow lens, red filter

Index

Printed in the United States
By Bookmasters